T0258766

Audio Mastering

Audio Mastering: Separating the Science from Fiction is an ideal guide to tangible development as a mastering engineer. This book offers a comprehensive overview of mastering processes, teaching the reader to link critical listening skills with analysis to comprehend the processing required to improve the quality of their audio.

Whether involved in game audio, composing for film and television, producing rock or programming EDM, the aesthetics required to manipulate and assess audio for mastering are a highly valuable addition to any artist or engineer's skill set. *Audio Mastering* encourages readers to engage in personal analysis and, in doing so, contextualise their evaluations by linking to appropriate and available processing tools. Discussion of processing conventions and theory are included to help readers create an effective audio chain utilising the best properties from each process, as well as practical examples and additional online resources to support development of skilled processing control.

This is a thorough and practical textbook for audio engineers, artists, producers and students on music production, music technology and music performance courses, and for aspiring and developing mastering engineers.

JP Braddock is a highly experienced audio engineer focused in both analogue and digital domain mastering. With over thirty years of experience in the music industry, he has delivered hundreds of albums over the last three decades, spanning a wide range of genres, including dance/electronic, folk, jazz and rock/metal. Through Formation Audio (https://mastering.ninja/), JP works as a mastering engineer, writer and educator delivering workshops for higher educational institutions across the UK.

Audio Mastering

Separating the Science from Fiction

JP Braddock

Routledge
Taylor & Francis Group

LONDON AND NEW YORK

Designed cover image: John Chapman

First published 2024
by Routledge
4 Park Square, Milton Park, Abingdon, Oxon OX14 4RN

and by Routledge
605 Third Avenue, New York, NY 10158

Routledge is an imprint of the Taylor & Francis Group, an informa business

© 2024 JP Braddock

The right of JP Braddock to be identified as author of this work has been asserted in accordance with sections 77 and 78 of the Copyright, Designs and Patents Act 1988.

All rights reserved. No part of this book may be reprinted or reproduced or utilised in any form or by any electronic, mechanical, or other means, now known or hereafter invented, including photocopying and recording, or in any information storage or retrieval system, without permission in writing from the publishers.

Trademark notice: Product or corporate names may be trademarks or registered trademarks, and are used only for identification and explanation without intent to infringe.

British Library Cataloguing-in-Publication Data
A catalogue record for this book is available from the British Library

Library of Congress Cataloging-in-Publication Data
Names: Braddock, John-Paul, author.
Title: Audio mastering : separating the science from fiction / JP Braddock.
Description: Abingdon, Oxon ; New York : Routledge, 2024. |
Includes bibliographical references and index.
Subjects: LCSH: Mastering (Sound recordings) | Sound recordings—Production and direction.
Classification: LCC ML3790 .B73 2024 (print) | LCC ML3790 (ebook) |
DDC 781.49—dc23/eng/20230829
LC record available at https://lccn.loc.gov/2023040281
LC ebook record available at https://lccn.loc.gov/2023040282

ISBN: 978-1-032-35903-8 (hbk)
ISBN: 978-1-032-35902-1 (pbk)
ISBN: 978-1-003-32925-1 (ebk)

DOI: 10.4324/9781003329251

Typeset in Optima
by codeMantra

Access the Support Material: www.routledge.com/9781032359021

This book is dedicated to all my family who have put up with me while I typed. I love you all dearly, and I promise to one day finish building the house. Though good frequencies just keep on getting in the way…

Contents

Contents

Acknowledgements

I would like to thank the publishers at Routledge and their team for the opportunity to explore my view on mastering as an artform in text, especially Hannah Rowe and Emily Tagg, your contributions have been invaluable in supporting the book's development. Equally to the reviewers, Russ Hepworth-Sawyer and Carlo Nardi, I greatly appreciated your pragmatic responses which have been very rewarding in developing clearer directives for concepts in the book. Lastly to my dear sister, your efforts and time questioning my thought processes are forever appreciated.

Introduction

The purpose of this book is to give the reader a comprehensive understanding of the mastering process. The skills required to master effectively are valid in all areas of music creation. Whether you are involved in games audio, a composer for film and television, or classical music production, the aesthetics required to manipulate and assess audio for mastering are a highly valuable addition to any engineer's skill set and comprehension of musical form. There are other genres like rock, pop, metal or country which are normally more associated with mastering as an intrinsic part of the production process, but the same principles apply. Learning how to engage in analysis and contextualise evaluations by linking to appropriate processing tools is required to achieve a mastered sound. Exploring processing conventions and theory can be used to create effective transfer paths utilising the best properties from each process. The transfer path is the signal path used to transfer the mix through processing that creates a mastered outcome. All these aspects are discussed in depth and practical examples are given to support the development of processing control, and to facilitate audio engineers to actually master audio effectively and achieve positive repeatable outcomes. Bridging the gap between perception of mastering as a 'dark art' and breaking myths surrounding what in truth is a technical and highly analytical process.

There are no 'presets' in mastering, and the skills and knowledge gained through developing this area can be equally applied to other parts of the production process. After all, the principles and perception of sound do not change just because recording or mixing is being conducted, but the directives towards observations and outcomes do.

DOI: 10.4324/9781003329251-1

From the source to the performance, at every stage in the production of music, each aspect when approached effectively actively contributes to the possible outcome that makes a 'great recording' that will be enjoyed for decades. This is truer now than ever with the sonic outcomes available. Quality of audio engineering tools are becoming more accessible from a financial perspective every year. After considering all this, to say 'mastering' can make or break a release is true, but to say mastering by some 'dark art' or unattainable equipment will make your music into something it is not or 'sound amazing' by just paying for this finalisation process is naive and profoundly misleading.

To understand this often misrepresented final sonic part of the production process, you need to remove prejudgments and misconceptions so you can engage in new learning with an open mind. I always point out to engineers I have mentored, who are well versed in engineering practice and theory, that the only aspect I possess as an engineer over them is years of listening to music critically. The knowledge, theory of tools and practical skill required to listen can be learnt. But comprehending what 'good' music sounds like is critical in understanding how to master. A valuable library of listening is essential to draw upon in any given listening context. Building your library of listening is something you will have to engage with on your own engineering journey. You may have already assembled much of this comprehension or are still developing your listening context. Either way, this book is designed to guide you in how and why mastering engineers engage in processing the way they do, and most importantly, 'why' their approach to listening and analysis differs to other parts of the production process. All this knowledge to be gained is transferable to any aspect of music making where aesthetic choices are made. After all, everything in audio can be distilled down into two simple aspects – phase (time) and amplitude (volume).

To start building this knowledge, we will explore these key areas:

- Audio referencing, learning to listen and good practice in approach to analysis.
- Making the link between tone and perception of amplitude.
- Contextualising analysis and its link to processing.
- Understanding concepts of momentary and static processing.
- Learning about tool type and processors, the link to primary and secondary outcomes.

- Learning dynamics and tonal balance, theory and practical approaches in a mastering context.
- Understanding commercial loudness.
- Mastering outputs and parts production.

There are online materials and worksheets to support your development that can be accessed via the following link: www.routledge.com/9781032359021.

Mastering audio is an art as much as any other niche in the production of music in its many varied forms. As will become apparent as this book progresses, mastering is not about a musical genre, but rather is sympathetic to the music presented, and is focused on skills to improve the sonic outcome. This can take many differing forms as there are many different and equally correct ways to go from A to B – 'A' being the mix and 'B' the master. But all the Bs will have commonalities in their presentation sonically, and it is this overview that is required to master any audio effectively.

This book conveys good practice and approaches to achieving these outcomes without micro focus on genre, but it also discusses clear directives in application in response to analysis. Genre is a distraction from the task at hand. This is not to dismiss it, but making the music supplied sound sonically improved in any method of playback in relation to the mix should be the mastering engineer's focus. This overview supersedes genre because the principles apply to all. Also remember that the source mix has been signed off by everyone when the outcome is sounding correct. The aim is not to produce or mix but to enhance the music in the context of the big picture. Mastering audio requires the engineer to look past bias in the musical genre and methods of production to see the sonic potential in any piece of music and to bring its best side forward, to make it 'fit for purpose' and to sonically balance it on all playback systems.

The directive in the following chapters is to teach you how to listen and analyse, how to evaluate your analysis and contextualise an appropriate link to processing, to teach you the methods of approach in application and to help you achieve your own transfer path based on your analysis of the source. Most importantly, it is to teach you how to evaluate outcome and know that the audio has been improved in all playback scenarios now and in the future.

What is mastering... really?

In its simplest sense, audio mastering is the link between the audio industry and the consumer. A more nuanced observation would be the link between the artistic process of song conception and its production ending in a mix and the delivery of the final mastered formats for the consumer.

To engage in any audio process, the scope of the inputs and outcomes must first be fully comprehended. Much of the work in creating an excellent master is in the efforts put into the previous parts of the production process. Without a great musician and performance, without a well-crafted song that is sonically well recorded and effectively mixed, the mastering process cannot deliver a superlative outcome.

If the mix source is flawed, mastering can only enhance what is there, not make it into something new. It would be better, but not amazing, which I feel should always be the focus to strive for, no matter the circumstances of the production process. A mastering engineer should always be supportive of other parts of the production process to achieve the best possible project outcomes.

As seen in figure 1.1, there is a transition in the focus of skills required for the differing parts of the production process. In the beginning, this is distinctly creative, and it ends with physical manufacturing or digital delivery, which are both highly technical. Mastering clearly sits at that end of the spectrum. This is not to say that mastering does not involve creative input, or that you do not need technical understanding to record an artist. Rather, the focus at the mastering stage is to technically be able to justify and rationalise each aspect of the signal path and what effect it has on the musical outcome – overall, more analysis, less experimentation.

DOI: 10.4324/9781003329251-2

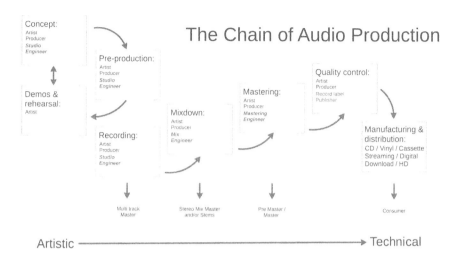

Figure 1.1 Production workflow diagram, from concept to the consumer

To master a song effectively and be able to replicate the process consistently day in and day out, you need to take off your creative, experimentation, production, songwriting hat and wear a technical/analytical mastering hat. This is one of the reasons why mix engineers across the generations have willing and wilfully passed their work on to the mastering engineer. It is a very difficult job to wear both hats for a given project, even more so as the artist/producer to record, mix and master.

Lastly, because of this difference in what is being observed, the mastering engineer is often viewed as a fresh set of ears on a project to bring those final enhancements without fundamentally changing the production. During mixing, the contact time with the music is normally over a very long duration, making it difficult to be objective or detached from the sonics created when moving on to finalising the audio. The mastering engineer brings this overview and objectivity to any project, no matter what the genre.

Goals of mastering

To achieve a positive result, the outcomes at play in reaching our ideal must be fully understood. Essentially, the most crucial goal should be making a master 'fit for purpose', meaning it will sound balanced on all playback systems and be appropriate for the restrictions of the delivery format.

Regardless of how a consumer listens to a song, it should sound good in their playback context. Often this is comparative when listening at home where they might just be playing a few albums or a playlist. Whether streaming or listening to a vinyl record, once the volume is set, they would not expect to have to adjust the volume or the tone. Equally in a club, they would not expect different tracks to sound widely different in tone or volume. In part, the DJ's job is to balance for loudness and in some aspects tone, but primarily they will have chosen a track because it sounds good in the first place – dynamically and tonally balanced.

If you are listening to the radio, you would not expect one song to be a lot duller than the next. This is partly due to the broadcast processor, a tailored signal path of processing for a given radio station. But mainly, it is because it is already tonally balanced due to the mastering process. Fundamentally a song requires a musically healthy dynamic range and tone in a genre to function in the real world, but again, the user should not have to adjust the volume or tone in playback of a single song or in back-to-back playlisting when listening to a genre. In any scenario of listening by the consumer, convenience is key.

When making music, an artist strives to achieve the highest quality outcomes in recording and mixing, even when trying to make sounds that are discordant or heavily distorted. The outcome may sound intense and broken, but these sounds still need the correct sonic qualities. Nasty noises still need bass middle and treble. This is just as true in the mastering stage when output would be a high definition (HD) outcome. At every stage, quality is essential in capture, processing and outputting.

The second aspect of the mastering engineer's scope is the master output format. To get a high quality, high definition (HD) file to compact disc (CD) quality, there has to be resolution lost from the music. The same is true restricting ourselves to lossy formats such as MPEG-1 Audio Layer 3 (MP3) or Advanced Audio Codec (AAC).There are better and worse ways to do this, but these are not immediately obvious, and artefacts sometimes only become apparent when played back on different systems. It is the mastering engineer's job to clearly understand these aspects of master delivery and continuously keep abreast of developments. These multiple outcomes or potential versions of a master are called the 'parts production' (for more information, see Chapter 16, 'Restrictions of delivery formats').

To make sure all these goals are correctly accounted for, it is important to have a logical approach to mastering workflow.

Mastering workflow

As previously indicated, the creative and technical difference between mixing and mastering can make for headaches if both parts are conflated as one. Another consideration is our exposure to or familiarity with the audio. In spending hours/days/weeks or even years recording and mixing a project, an engineer becomes intensely intimate with every micro detail in each mix, and equally all its remaining flaws are present in mind. A mix is hard to finish; part of the skill is knowing when parts are just being moved around and the changes are no longer improving the outcome. This closeness, and often emotional entanglement, means it is difficult to be objective in our analysis when it comes to the overview required in mastering. Often the need to 'fix' errors in the mix are brought to the mastering, meaning the song just becomes over processed. Hence there is the need for the fresh ear of the mastering engineer and clear objectivity around what processing needs to be applied.

If you have mixed a lot, I am sure you have experienced the feeling of 'what was I thinking?' after working into the night and listening to a mix fresh the next day. In your first listen, everything that needs adjusting seems immediately obvious. This freshness in change of perspective is what the mastering engineer brings each and every time, but with a honed sense of objectivity. Looking back to the production workflow diagram (figure 1.1), a separate recording and mix engineer provides neutrality and a fresh objective pair of ears as the product develops. It is the producer's role to have the overview and vision across the whole production process to work with each engineer and the artist to bring the project to fruition.

In the last few decades, and probably even more so as technology strides on, the delineations between these roles have become increasingly blurred or even to the point that there are none. The artist is the creator, producer, recordist, mixer and mastering engineer. This is not a negative aspect; in my opinion, technology and its development should be embraced by all in the artistic community. It opens doors, breaks down barriers to new experimentation and methods of connecting or sharing our art with the consumer. What is helpful, no matter what the makeup of the project, is clear demarcations in the production process, allowing for an informative way to evaluate a project's development. History can always provide lessons, and there is a wealth of history of music production to be absorbed. To effectively use this knowledge, the correct approach is needed to enable analysis.

Enabling analysis

To manipulate the audio signal effectively, it is critical to be able to perceive errors in the source and comprehend the aesthetic changes considered appropriate for a commercial outcome. Without being able to perceive the potential issues when listening, it is impossible to make an informed decision as to how the audio needs to be manipulated. Simply, if it cannot be heard, it cannot be changed or evaluated. In the latter case, any changes actioned are unlikely to improve the outcome, or will just make it different or worse, or at best, it will maintain the status quo.

One of the aspects that most informs a mastering engineer in their work is their sense of perspective built up over thousands of hours of listening in critical spaces. If researching mastering engineers' working practices, you would find they all consistently work in the same environments with the same monitoring path. Any changes to their working environment are highly considered beforehand and upgrades are applied only after in-depth research and practical testing have been undertaken. This is in contrast to some artists and producers who crave new environments and signal paths to impart fresh new atmospheres and sounds for their productions.

This consideration of an emotive and analytical response to the perception of sound can be viewed in the transition from the beginning of the creative process to its ending during mastering, the latter being highly ordered and analytical, requiring a consistent listening environment and path. This contrasts with a more hedonistic or free approach in creating music, when loud sounds good and makes us feel good and quiet feels soft and relaxing. But different genres offer other perspectives in the creative process. Overall one thing that should be clear to everyone involved in music making is that there is not a definitive approach or singularly correct outcome. This

DOI: 10.4324/9781003329251-3

principle of transition in the perspective across the production process can be observed in figure 1.1 in Chapter 1.

Considering the above, a question presents itself, 'How do engineers effectively move between environments?' or 'How does a mastering engineer know that what they are perceiving is indeed the outcome they think they are hearing?'

The answer is audio referencing.

Audio referencing

Music making often references previous creative outcomes. Whether consciously or unconsciously, creative people absorb a great deal from the music around them and from the soundscapes of everyday life. This osmosis of musical form is an audio engineer's library of sonic reference.

Simply listening to music is the best way to improve your ability to record, mix and master. In many conversations with my peers over the years, I received one consistent piece of advice, to critically listen to as much music as you can. When I discuss audio with new generations of audio engineers, I feel it useful to point out that the only tangible aspect I possess as an engineer in comparison to them is my sense of perspective from my library of listening built from constantly referencing audio in critical spaces. It is not some technical prowess, critical insight or knowledge; there are no secret skills or dark arts in audio, all theory can be learnt and skills practised. I have just listened to a lot of music and developed a more sophisticated sense of average across genres to what quality is from these decades of absorption.

The nature of familiarisation with excellent sounding music will enable you to comprehend what makes a great sounding track, not just the music itself, but its tone and dynamic – the sonics. This sense of overview is critical. A song that translates in all environments, whether played in a venue, mobile device, massive club system or a beautiful sounding hi-fi, is a song that has the composite parts of balance required in both spatial and stereo impression, tone and dynamic, thus making it a great reference. The spatial is the ambience and dimensional depth or often referred to in mixing as the depth of field. The stereo impression is where instruments are panned in the stereo image.

I'm not overlooking the roles song writing and voicing have to play in the creation of this positive sound balance. If all the instrument registers pocket in frequency and the riffs interplay and wrap around each other, then there is a natural density and balance within the song's construction. This gives a positive balance to the mix and hence the master. A great sounding outcome has to have the correct humble beginnings in the register of the parts. If they clash, it is never going to have clarity or loudness. It will be masked and overly weighted.

References are highly personal; what makes great music is definitely in the 'ear' of the beholder. But the more genres and eras you digest, the more you will see that good tonal and dynamic balance is a consistent average throughout quality outcomes. Decades go through fashions in dynamic range and tone that are normally driven by the technology of the time and its use artistically. But when observing the music in overview, this premise of tone and dynamic balance holds true.

In my opinion, a reference should be a piece of music that 'you' feel sounds superlative wherever or however you listen to it – a great musical and sonic translation on all systems.

In listening and analysing music, it is critical to be aware of the most important aspect of our perception of audio – loudness.

Loudness

We live in a digital age. If you have grown up making music primarily in this realm, your focus will undoubtedly have been on the peak of the signal. Peak metering is commonplace on digital systems, because it stops what none of us generally want (unless creatively) – digital clipping.

At best, it adds harshness to the audio, but generally it just sounds down-right nasty and unmusical. As the peak is the constant observation, often tethered with a 'peak hold' function on a visual meter scale, it can be easy to forget or not to have observed that the peak of any signal is not the impression in our perception of loudness. The average level, often denoted as the root mean square (RMS) is. This is a metering measure that has a window of approximately 300 ms, though this can be adjusted on some metering systems. The window moves along with the audio in playback to give an average measure for the value of the waveform overtime in decibels (dB). This average level reading is generally noticeably below the peak reading in dB, that being the maximum amplitude the audio reaches at any given moment in time in samples. Often this is noted as a 'sample peak' meter.

With differing metering systems, you may also observe peak averages (hold) and RMS averages or RMS peak hold measures. These should not be confused with the simplest measurement of peak/RMS. Peak for the maximum value and RMS for the average power or the musical loudness. I will expand on these principles as the book progresses.

Imagine observing a sample peak meter, with one sample on the audio timeline normalised to 0dBFS (decibels full scale). Playing across this single sample, the meter would show a peak at 0dBFS and may audibly create a click. Though if this signal is observed on an RMS meter, nothing would be measured. Coping that one sample 44099 times back to back to make a second of audio at 0dBFS and then pressing play, the peak meter would still read 0dBFS the same as before, but the RMS meter would now also evidence a response. This is because the signal now has sustain in the envelope, making the display evidence a continuous average level on the meter. This RMS measurement is how sustain, i.e., loudness, is distinguished on our visual scale. I would turn your monitor level right down before doing this practically, because the outcome heard will be very loud indeed in comparison to the single sample.

Perceived loudness is measured using an RMS meter, and the maximum level of any one given transient with a sample peak meter. The latter does not tell us anything about how loud it is, the RMS meter does, as would a loudness unit (LU) meter, with the focus on the average level of the signal over time and not the individual peaks. An LU meter is normally used in conjunction with a True Peak (TP) meter to display the potential peak of the signal. I expand on the use of TP and LU in later chapters.

Comprehending the envelope shape linked directly to the impression of loudness is the key to understanding dynamics and how to manipulate any dynamic range change effectively. Being able to observe both transient material (peak) in music and sustain (RMS) enables engineers to achieve targeted dynamics in differing areas of the music's waveform.

To analyse productivity, there is a clear need to observe equal loudness contours.

Equal loudness contours/curves

Our ears are our best friends and our worst enemy. Not adhering to equal loudness means decisions in the end will be biased by differences in loudness and not the actual differences in the sonic. Equal loudness means that when comparing one audio aspect to another, or even the same audio, there should be no difference in loudness between them. This is one of the reasons why a compressor has a 'makeup gain', in compressing a signal it is inherently quieter. Hence there is a need to rebalance the output (makeup) to match back in loudness to the input level. Thus when bypassing the compressor, the comparison to process is at equal loudness. This also maintains quality in the signal path as the headroom and signal to noise ratio is the same at input as output (I/O). Critically the change then observed is the authentic change in dynamics. This is the actual point of using the compressor in the first place.

There is an easy experiment to prove this psycho acoustic effect in loudness difference.

Take a reference track you know well, copy it to facilitate an A/B comparison in your DAW. In switching between the two, soloing each in turn there will be no perceived difference. This is correct because it is clearly the same audio, a copy of itself. But, if you now turn one of the tracks down by a dB, and make the same listening comparison, what do you hear?

Firstly, a difference. Not just the obvious perception that it is quieter, but a tonal difference as well. There is no difference in the actual audio because it is the same source, but there is clearly a perceived change. The quieter version sounds weaker, less full. Or the louder sounds brighter, but with more power in the bass end, a smiley face EQ curve difference – a power curve. The stereo impression (width) is also likely to change with the louder sounding wider comparatively. Knowing there is no physical difference and our ears do not lie, it can only be our perception of the audio that is changing. Our ears are being fooled by a difference in loudness. Though what is being perceived is correct, loudness changes our perception of the source even if the audio is identical except the amplitude difference.

This obviously has many implications, if for instance boosting an EQ, it will increase amplitude as well as change tonal balance. Turning this EQ boost on/off and observing the EQ plus the loudness difference, makes the sound with the EQ perceptively better as it is louder, sounding fuller and more impactful.

Turn down the EQ'd version to rebalance the EQ output and achieve equal loudness. Now the difference between the EQ and the original signal can truly be perceived. The opposite is observed when cutting the EQ and bypass on/off – the EQ'd sound will be weaker or lacklustre, and with the EQ bypass, the volume difference will be louder, meaning it probably sounds better without the cut. But the output simply needs rebalancing with a gain increase to match loudness in the same way, as when comparing compression outcomes using the makeup gain. This issue of loudness perception is one of the reasons why most mastering EQs have 'output volume' control or a set of 'trims'.

By the same token, in applying a limiter boosting the outcome by 1 dB, we already know whether it will sound better with or without the limiter. Without having to test, loudness dictates that the limited version will sound better because it is perceptively louder – it will sound better in this flawed comparison. You can hear the true effect of the limiter by achieving equal loudness when turning the limited version down by 1 dB at its output. If limiting harder than this, the perceived correct levelling will probably be less than the gain reduction because of the limiter's effect on the tone of the audio.

You should match the loudness levels in all comparisons. In the 1930s, Fletcher and Munson [1] conducted a set of experiments with loudness and human perception of frequency balance; this work has been corroborated by many others who fundamentally reached the same conclusions. The long and short of this is that there is an optimum listening level for audio that is dependent on environmental factors, a sound pressure level (SPL) of between 75 to 85dB where perception of tonal and ability to discern detail is the most transparent. In adjusting the loudness around this, our perception of tone changes, but loudness is subjective and there are many differing opinions about what is correct. You may see this ideal listening level referred to at 86, 85, 83, 80, 75dB in the stereo perspective or is suggested as broadly between 75 to 85dB. But I have found from many years of experimentation in the listening context of mastering that the sweet spot average level of 83dB SPL is optimum in a controlled environment. This assumes the use of a mid-field monitor system which is normal in a mastering context. This is where the monitors are placed from approximately 2 to 4 metres from the listening position, near field would be 1 to 2 metres and far field past 4 metres.

Many of what would seem to be discrepancies in the 'correct' level come from differing interpretations of the relationship between SPL, loudness and reference levels. The latter dictates the ideal headroom of the audio signal path and is normally linked to a metering scale. As you will see as these ideas are explored in more depth, it is also helpful to link SPL to our visual meter. Hence differences in how this information are presented because the scales are different. SPL is measurable to a scale in dB, but loudness is subjective. The way loudness is perceived changes with the relationship of how it is observed. For example a piece of music going from quiet to loud sounds like a bigger change than when it is adjusted by the same amount from loud to quiet. This phenomenon of psycho acoustics can be observed in our loudness difference A/B test that I hope you have conducted with the small loudness difference in place between the same sources A and B of 1dB. In observing the perceived change from A to B, the effect of the loudness difference is fuller and brighter and is heard more so when changing from A to B and less so when changing from B to A, assuming A is quieter relative to B being louder.

As a theme, if a sound is turned down, it sounds less bright and less full when observing this in a critical environment and when listening to complex full range music like an audio reference. You will notice the audio not only becomes less bright, but it also moves backward into the speakers and the spatial impression loses transparency and width in the stereo impression. Turn the audio back up, and the audio will move forward out the speakers, becoming brighter, fuller and more transparent and wider. Obviously, the actual audio file is not changing, but our perception of the audio clearly is.

This is a straightforward test to conduct and observe as long as it is conducted critically. What you will also notice as the audio gets louder past the suggested range of 75–85dB SPL threshold, although it feels more powerful, is that it starts to become less transparent, more compressed sounding, and the transients are less defined and the bass is harder to hear around. In listening to these aspects carefully, you can find the optimum playback level without the support of an SPL meter. It is just about listening for where our audio reference sounds the most transparent, and where you can hear the most detail and depth and width in the music observed. Once you find it, place an SPL meter in the sweet spot (assuming that is where you would be listening, because why wouldn't you?) and it will read about 83dB SPL C-weighted (dBC) on average give or take a dB or two. Again this assumes you are conducting the test with a midfield monitor system in an acoustically controlled environment. What you must do in your environment is trust your ears during this listening test for where the music sounds the most transparent. At this level, the SPL measured is correct for your environment. The numbers are a guide, your ears are correct. The more controlled the environment, the more this ideal of 83dB will be achieved; the less acoustically controlled the environment is, the lower that likely SPL reading will be. This principle is explored in-depth in the following chapter.

But in many regards, the important outcome of these experiments is simply that louder sounds better. From an appreciation of audio playback, this does mean a more compressed audio master makes for a louder comparative in playback when not rebalancing for equal loudness. Thus, making the audio sound better in perception relative, a less compressed master will be relatively quieter. This comparison is obviously bias by loudness difference, and the listeners do not hear the actual sonic difference between the audio. They are hearing the difference plus the loudness effect in their perception, something consumers can be influenced by daily in any digestion of any audio playback. This is a fundamental issue behind what has been referred to in the music press over many decades as 'The Loudness Wars'.

Louder sounds better

Louder sounds are perceptibly better if making an ill-informed comparison. This is how most consumers listen to music every day. If you are a consumer, you simply want to play the song. You are not considering its dynamic range, you just expect the song to play, and by the nature of human perception, louder tracks will sound better comparatively in any playback scenario where the listening level is simply not just too loud.

This phenomenon has been affecting consumers' choices from the earliest days of modern commercial music. In the 1940s the use of the jukebox was a popular way to enjoy music socially. Singles were played back on these fixed volume players that differed in output volume depending on the loudness of the cut. Record companies noticed that some records had a better consumer response than others because of the power in playback – it sounded better! This led to the position of 'transfer engineers', the forerunners to mastering engineers who were asked to make the 'transfer' between the tape and vinyl louder. This was achieved utilising compression, the new technology of the time, but at first by equalisation of tonal balance. This was the possible start of the loudness wars. The louder vinyl's perceptively sounded better, as they were louder in playback, but this did not mean necessarily they always sounded better at equal loudness. But the compressed audio on a louder transfer will have achieved a more effective cut, therefore when making a comparison back to its unprocessed source at equal loudness, the compressed vinyl in principle should have sounded better – a good outcome from an engineering point of view.

Fast forward four decades to the advent of digital tools and digital delivery. The invention of the compact disc (CD) meant there was now a different focus on the signal. With previous analogue formats, the observation in level was around the RMS to maintain optimum level between noise floor, saturation and distortion. Digital audio delivery meant this focus shifted to include the peak to avoid digital clipping. But it is the RMS that gives our overall perception of volume or density to the sound. With the creation of digital tools, it was possible to increase perceived volume by simply reducing the peaks and pushing the level to 0dBFS, making the focus change from the creation of positive density in the RMS to also achieving peak reduction, increasing the perceived loudness in the final consumer CD.

At first, this was straightforward wave editing to reduce the larger peaks. This is quite a passive process as the volume is being reduced at that moment in time and is not actually changing the shape of the transient, just reducing in amplitude around the zero-axis crossing point. This can be very time consuming; digital limiting made the process in real time but then shaped the transient, and thus changed the perspective of the mix. A little limiting, skimming off the highest peaks has minimal effect, and a lot of limiting significantly changes the overall impact of a track. On one hand, this can be seen as an ideal outcome as the perceived loudness is increased, but equally as a destructive and unnecessary manipulation of the audio. This digital manipulation for loudness was something that affected almost every aspect of modern music making. Hence the perception of the loudness war continued unabated onward and upward in perceived loudness and reduction in dynamic range.

Jumping forward another three decades and now music delivery is moving into a world of loudness normalisation where the average level of a digital audio file determines its playback level on a given system. This is the start of something good for the resurrection of focus on the RMS and dynamic range, and more so for returning to the principles of the analogue realm where focus was solidly on RMS for perspective in loudness. But not all consumers have access to or are using loudness normalisation, meaning a master must work effectively at both peak comparison and under loudness normalisation relative to other music outcomes. I discuss how to achieve this in Chapter 14, 'Navigating loudness'.

I should point out at this juncture that loudness and compression are not bad things – a well mastered song should be balanced in loudness to have an effective dynamic range in all playback environments whether these are noisy or quiet. Some genres need a clipped or limited peak to sound smooth and effective in this translation. The mastering engineer focuses on effective control of the dynamic range of the music to improve the listening experience for the consumer. The impression that there is a war afoot is often viewed as misguided or ill-informed by those actually finalising the audio. There are many examples of people looking at waveforms comparing the 'negative' effects of loudness. But as mastering engineers know, listening to the music is the key, while looking at a meter or the waveform is a distraction. There are always examples of poorly translated music from a mastering perspective, but do not forget the millions of other great sounding musical examples in your narrative.

How this develops in future depends on many factors, but one thing will always hold true in comparison from a mastering perspective – a song that is louder than another by any method of measured comparison will sound better perceptively. A true comparison of any audio at equal loudness informs an engineer of how it actually sounds different. In making this comparison from mix to master, the master must sound better than the mix in all regards. If all these outcomes are true, the master must be better than the source no matter what the method of measured loudness is. How it sounds at equal loudness is the only truly useful factor in the end in critical comparison.

Generation lossy

The positive outcome of the technology implemented at the start of the millennium enabled a generation to open up the music industry, making music creation and distribution available to all, especially those outside of the previous fixed industry framework of record labels and physical distribution. This also brought about a reduction in the quality of how consumers digested music on a daily basis. Lossy audio has several negative factors on our perception of audio (how to isolate and distinguish lossy are discussed practically in the supplementary material available online, see the introduction for the link).

If you are part of this generation, your journey to hearing how music actually sounds lies in part in stopping listening to lossy in all its forms whenever being critical. Which if you intend to master, should be most if not all of the time!

My generation and previous generations of sound engineers had the advantage of listening only with lossless media. It may have had a little hiss and/or pops and crackle, but it did not have any of the negative degraded attributes of lossy audio. Also, as a theme, people listened to speakers a lot, but headphones have become more of a mainstay as the decades and technology have progressed. The Sony Walkman launched mass portable headphone use, and this has been built upon since with iPod and other battery operated portable devices. They have all changed how consumers digest music from day to day. I have canvassed students about this over the years and many do not own speakers. When I was in my twenties, everyone I knew had speakers of some form. Because listening habits change for many rationales, it is important for engineers to be conscious of how our output is being consumed.

Moving forward, music delivery is now coming into an era of lossless audio as most streaming platforms offer lossless playback as an option of service, and those that still do not say they are close to launching lossless options. Aside from the general use of headphones in gaming and virtual reality (VR), the ability to use binaural playback to create the impression of a 3D spatial audio environment is a centrepiece in audio gaming design. Mastered music is inserted into this environment, though normally it is the visual aspects that are in motion in the sound stage and the music stays in our stereo impression. Generally music spinning around makes it difficult to listen to even if it is just used as the background to a scenario played out visually in a game. There is a difference to this foundation in the use of music in media and the directive to bring spatial audio via Dolby Atmos to music only listening, something that has been tried in varied guises in the past. This is expanded on in Chapter 16 'Restrictions of delivery formats'.

Ear training

Ear training is the simple act of listening to music when being critical in observation, i.e., listening with a considered monitor path in an appropriate environment. The more music you digest critically, the wider your understanding of genre and the bigger impact on your comprehension of what quality is and how a finished product actually sounds. Many mastering engineers talk about their formative years working on transfers and on quality control (QC). With years of listening to other mastering engineers' output, you develop a sense of quality and perspective for correct tonal balance and dynamic. Most mastering engineers would say they spent a decade listening before being in a true position to fully comprehend this aspect.

Gone are the days of the mastering studio being able to offer decade-long apprenticeships to trainee engineers in the studio; even in my generation, this was nearly dead because our industry is ever evolving. In my case as a producer/engineering in our studio, taking that work to commercial mastering environments was the catalyst for my interest in the field. I found intriguing the technical nature of engagement with sound and the small balances made to achieve a big difference in the general fitness of playback, and this mostly revolved around being truly critical while listening. It was over a decade before I found that sense of quality and perspective, my library of listening and the confidence in output to call myself a mastering engineer.

It is also worth noting that whenever or however you may listen to music, this still builds into a sense of average and your library of listening, especially if the music observed has also been assessed in a critical controlled environment leading to comprehension about its translation in the real world. Coming back to my first statement, 'the simple act of listening to music is ear training' is true no matter how or where music is consumed.

Hearing dynamics

One of the most important aspects in starting to be able to critically listen effectively is in developing the link between volume and tonal change. The principle of 'louder sounds better' is born out in every aspect of our listening. When mixing a given track, if you adjust the fader to address where the vocal sits relative to the other instrumentation, you are also at that point adjusting its tone relative to everything else. The louder it is, the more brightness and relative weight it will have perceptively. When quieter, it will dull and reduce its punch. Hence when you turn up the fader, the vocal comes forward in the mix, sounding on top of other aspects. The same is true if the vocal is changing dynamically, after all, dynamic range is just the travel of amplitude over time. Hearing tonal change in the dynamic of the vocal when assessing a finished mix, means there must be an amplitude change occurring. From a mastering perspective, generally it means there is too much movement in that aspect, or overall in the sound balance of the vocal across the song.

Becoming fully aware of this relationship between amplitude and tone makes the link to processing control come into focus whether you are mixing or mastering. But to start to perceive these changes in volume correctly, it is critical to observe these differences in an effective listening environment, meaning time must be invested in understanding the monitor path.

The monitor path

To facilitate our ability to analyse correctly, the following three aspects need to be carefully considered:

the listening level
the domain transfer
the playback monitoring and environment.

These aspects come together as an audio playback chain referred to as the monitor path. To remain truly evaluative and critical at all times, this path must stay consistent throughout the entire mastering process. This informed listening builds a context to our perception and library of listening over time. For instance, if we changed the speakers wholesale while mastering a song, our perception will be tainted because our impression of the audio will change and our listening context is not being developed.

Fixed listening level

The magic number is 83dB in my opinion. This is a measurement of sound pressure level (SPL) in dB, not to be confused with other ratings of energy/power or perceived level sometimes expressed in dB. In straightforward terms, there is a relationship in ratio between all these aspects, but it is the sound level coming out of the speakers in the central position that is of concern – the sweet spot. As discussed in the previous chapter, Fletcher and Munson [1] carried out experiments with our perception of frequency and listening level in the 1930s. By conducting the suggested tests in your own

DOI: 10.4324/9781003329251-4

environment, you should now be aware that louder sounds appear fuller, while quieter sounds are duller and lack energy in comparison.

Logically, if this change in perception exists, there must be a point where the listening level sounds the clearest or best defined. Broadly there is a consensus this is between 75 and 85dB SPL. However, if the music is turned up, it will sound better in principle as per our previous test. In adjusting our monitor level to the point our reference track sounds the clearest or most transparent, it will present the most detail and definition in the music observed. When then measuring the SPL in the stereo listening position, it will be around 83dBC, give or take a dB and/or the accuracy of the SPL meter and monitor environment. C-weighting (dBC) is best for measuring musical loudness where bass is an important part of the perspective in sound level. Do not use A-weighting (dBA) as the impression of the bass power will be significantly reduced. If the level is pushed louder than 83dB, it will be perceived as if the music sounds more compressed – more impactful, but not as transparent. This is often most noticeable with transients on the percussive elements which start to become less clear or smeared in comparison to the lower level moving back towards 83dB. Even though the music is more powerful above 83dB, it is not as defined. It is the intelligibility that needs to be listened for, not the musical power in locating this ideal listening level.

This experiment has been conducted for years with many different cohorts of engineers. Without fail, they always find the optimum listening level when critically focused in an effective environment; this means in a near/ midfield acoustically treated room. Once found, the SPL meter positioned in the sweet spot reads an average of around 83dBC during stereo playback. I can detect that level within a few seconds of listening to anything in a controlled studio environment because it is so familiar.

If you're testing in a large cinematic environment in a much larger room, you would find the music would probably sound best at a higher average of 85dB or more because the music is more diffused by the room (even though treated) and there is less direct signal. This means more power is needed to perceive the transparency of the source effectively. Conversely in a smaller room with a near field speaker, you will find that 83dB SPL is a little loud and you are more likely to settle at 78–82dB when listening for your reference's transparency. There are more reflections and the loudness feels more intense relative, hence transparency is best heard with a little less level overall. Either way depending on the environment and monitor system, the measured SPL will be within a few dB of 83. As low as 75dB always feels more observational, listening to music for music's enjoyment, like enjoying the sound of a good Hi-Fi system, but not helpful for critical listening, as the sound lacks directness in the transients and I would consider it soft in this regard.

Loudness is subjective to the listeners mood, but it is also affected by the environment and monitoring system's impact on our perception of the music. You need to experiment with this knowledge in mind and really listen for the loudness that makes the music in playback sound best and clearest, most transparent in your environment. That is your ideal. Measurement is numbers. Our hearing is our key to the best possible outcomes and you must start to trust what you hear, not what you are told is correct. Information and data are only guides. Comprehension of all these factors enables us to utilise our hearing in the most effective way.

Try it yourself, and you will likely repeat the outcomes observed by Fletcher and Munson [1], and in doing so learn that there is an optimum listening level in assessing music critically. Though if you're responding purely emotionally, louder will always sound better as we want to 'feel' or 'enjoy' the power of the music. But this is not our job as a mastering engineer; our need is to be highly critical, analytical and comparative at all times to avoid being fooled by loudness. So how do audio engineers maintain this accuracy in listening level day to day? They work with calibrated monitoring.

Calibrated monitoring

To maintain equal loudness with a consistent listening level, most professional monitor controllers are stepped. This means the control has fixed intervals in dB, and as with many mastering tools this improves accuracy. It is intrinsic to circuit design to achieve the most effective and accurate adjustment of the signal path possible. With the appropriate approach, it is possible to create calibration levels for non-stepped monitor controls. Either way, the important aspect is being confident in your listening level.

Calibration, if used appropriately, allows engineers to share their evaluations of the audio with other engineers in different studios around the world when they are working to the same calibrated level. Everyone is listening in the same way and can discuss the audio and its critique with a common outcome in terms of perception.

Equally it means a common headroom can be dedicated and used by everybody involved in the project, which is a valuable outcome in the real world of commercial audio. The most important positive of maintaining constant perception in listening level is that it makes all the music listened to comparable on the same footing. All the music worked with and references used fall into this equal loudness mixing pot of perspective in observation of the sonics. Over time, this average will help you to achieve a heightened engagement with critical analysis and observe what quality actually sounds like. This also makes it obvious when you are not listening to it!

Whether a stepped monitor controller is available or not, to actually calibrate a system it is vital to have metering.

Metering

This brings us full circle back to the need for a consistent observation of level. Most modern digital meters allow the observation of peaks and RMS. This could be full scale dBFS, or a relative scale dBr, in which 0dB could be anywhere on the scale, but peak measurement is still actually identical. When observing a meter, it is important to remember that dBs measured stay the same, but the graphics and scales can change making it 'look' different. The actual background measurement of peak and RMS is consistent unless the way this is measured is changed. When considering what the meter is informing us about the sound, peak is just that, a measurement of the highest amplitude of the signal at a given moment in time. Just because a level is observed peaking at 0dBFS, it does not mean it is loud. Loudness comes from the average weight of a signal, the RMS. Imagining a long deep snare sound and reducing the release/sustain, the peak would remain equivalent when viewing the meter, but the power or weight of the sound would reduce. The RMS level would drop on the meter. In reducing the sustain and release further, the initial transient would only be a few samples and the peak meter would still be reading the same, but audibly, the sound would now be a small tick or click. With no body or power, there would be no RMS on the meter as there is no sustain or power to the sound. This shows that sustain is crucial to perceived loudness; the more a sound has, the denser and more powerful the outcome. Bass guitar is a good example. A long bass note will have power, a muted bass will have impact but only momentarily. This is borne out on the meter in measurement of RMS, a sustained bass will have a high reading, the muted bass note will have little impact on the meter until it is played repetitively, then the potential sustain increases over time making the RMS reading on the meter rise.

RMS delivers the loudness impression on a meter, peak is just the maximum transient of the audio, not the loudest section of the audio. Now that this is established, considering the RMS of the signal and not the peak is critical in making any audio perceptively loud. Before digital audio, this was the main focus of any audio engineer's observation, as the RMS delivers the power, a peak is just that. On a high quality analogue system, the peak will just saturate a little even if it is quite powerful, but if the RMS is too hot, the peaks will potentially be overly saturated or noticeably distorted. The sound is no longer 'clean', it has been driven. If the RMS is too low, there would be a relative increase of noise present in the system because the average level is closer to the noise floor which exists in any analogue system. Since digital has become the norm, generations of engineers have grown up observing peaks because digital clipping is generally nasty, producing abrasive negative distortion outcomes, hence the explicit focus on dBFS peak metering in DAWs. But to make our audio sound dense, the observation needs to be on what is happening with the RMS/average, the power/loudness of the signal.

Bring this back around to calibration and measuring the loudness relative to a listening level. It does not matter what metering system is used or who manufactured it, it just needs to be able to read the RMS of the signal effectively. This means the scale and range of the meter works accurately in the relative area of use. For example, if the scale viewable on the meter is 100dB, and the area to be read is around -20dBFS, the visual is only showing about 10% of the viewable area required in our focus. In changing the scale to 40dB, visually the reading area in view would now be more than twice the resolution. It is important to be able to read the range on the meter required accurately and effectively, or you are just making the job unnecessarily difficult.

At this point it is worth noting standards in reference levels for equipment. This has always helped in interfacing between different processors in the signal path coming from the days of analogue where it was critical to maintain headroom to avoid distortion and adding unnecessary noise. This is always linked to metering, but in a digital environment the same rules do not apply because headroom is almost infinite as long as a digital clip is avoided. But this does not mean it is not a good idea to adopt these standards, as the principle of observing headroom means the focus is on the RMS/loudness and there is not the need to worry about clipping and overload because there is enough headroom dictated on the system. These reference levels are normally expressed in dBu, which are dB units linked directly to voltage. Analogue systems use dBu to define the correct operating voltage for equipment in use. The European Broadcast Union (EBU) standard alignment is 0dBu, equal to -18dBFS [2]. In America, the Society of Motion Picture and Television Engineers (SMPTE) define this operating headroom as +4dBu, equal to -20dBFS [3], which gives a larger headroom overall. This means an analogue mixer would operate at the optimum average/RMS level around this 0dBu or +4dBu, giving the system a headroom before actually clipping of approximately +24dBu. In working at the correct operating level, headroom will be maintained throughout production, and equipment can interface at the correct operating voltage. Taking this thought process across to digital in dBFS to be equivalent, the average level would need to be set at somewhere between -24 and -18dBFS to maintain the same principle of focus on the average level and not worry about peak outcome.

For those of you who are new to metering with focus on the RMS, the in-complex K-system meter is available and built in as a scale on many DAW's metering systems. This is a great educational meter scaling introduced by Bob Katz that brings three different commercial styles of metering systems into one easy-to-comprehend system. This will allow you to dictate a headroom in dB on the system, putting your focus on the RMS and not purely on the digital peak. There are plenty of third-party plugins that will help you get started with this scaling. This is not something I would use during mastering because the scales are restricted in scope, but I have found it is very helpful to those new to observing RMS for loudness when engaging in the principles of calibrating a monitor system. Bob Katz has published an informative Audio Engineering Society (AES) paper [4] if you require further insights as a starting point.

With clearly denoted scalings and the ability to change the decibels relative to reference level (dBr) centre, it is helpful to start to observe dynamic range in a digital metering construct as opposed to how many engineers discuss outcome incorrectly and are focused on peak thinking that leaving a 'headroom' from the peak level to 0dBFS in digital is the important factor. I will explain why this is not the case in developing a broader understanding of resolution and the mastering process. Our focus clearly needs to be on the RMS and not the peak.

These differing reference levels, metering scaling and the measure of SPL in relationship can be confusing because these measurements in different publications do not always tally. In starting to calibrate your own monitoring, I would not worry about these discrepancies but would focus on the good working practice evidenced by the principles of linking our operating level, system headroom, metering and monitor level SPL together to achieve a constant listening level and the ability to change headroom in the system – all this while maintaining quality and the same fixed listening level for focusing on equal loudness.

This brings us all the way back to the beginning of why it is important to engage in all this. Ideal listening level means adhering to loudness curves, while fixed listening level means operating at equal loudness always. Both of these mean adherence to best practise with the headroom of the systems in use maintaining the best quality possible. Those are the principles that should be focused on making our working day easier and more controllable.

What is really required to calibrate and work day to day is the most responsive set of meters relative to the signal path. I personally find it helpful to be able to monitor any key change in the signal path, i.e., transfer path digital to analogue converter (DAC) and analogue to digital converter (ADC), as well as monitor path DAC and any digital interfacing change within hardware. This is in addition to the final output stages of the audio relative to the 'print'. This term means the audio is being played and captured back in real time, printed or to print the audio. This is different to a render, which is an offline process conducted inside a DAW which could be real time or not depending on the method of bounce.

To calibrate practically there needs to be a fixed level in RMS playback, and this level needs to be locked to an amplitude value on the monitor control level. All measured in dB, meaning one can be adjusted accurately against the other maintaining listening level during any changes in relative headroom. The best method in approach is to use pink noise levelled on the RMS meter leaving a large headroom before peak – the top of the system. Pink noise is a broadband noise with more perceived power in the lower frequencies as there is less in the highs. This is the most musical sounding noise in its distribution of frequencies. If you summed all the music in the world together, it would probably sound a lot like pink noise. I have experimented with creating my own noise curves, but it is going around the houses somewhat in the end. I concluded that pink noise works very well for this task in multiple musical listening scenarios.

Setting an RMS to peak range of between 24dB and 18dB is a good area to aim for as this achieves a large potential dynamic range to work in and peak clipping is highly unlikely with mixed audio when averaging to the

RMS level set. With the pink noise averaging on the meter in the digital system at our decided RMS level, observe the level in SPL in the studio's monitoring sweet spot using an SPL meter, this should be set C-weighted. As with using pink noise, a C-weighting (dBC) allows for a more musical appraisal of the bass frequency relative to the treble. Adjust the monitor level until the SPL meter is reading 83dBC, then make note of the stepped value in dB on the monitor controller, or if it is a variable control, just mark with a chinagraph pencil or tape for later recallability.

To be clear, measure the SPL in the sweet spot with both speakers outputting. You may see this same calibration discussed as measuring 83dB at -20dBFS or 85dB at -18dBFS, which are the same thing when referencing 0dBu in analogue depending on the meter reference level calibration. This is also a standard room calibration for film work set out in Dolby/SMPTE standards [5]. This cinematic scenario is inherently more dynamic than audio mastering and is conducted in a larger studio environment than a more acoustically controlled mastering room. Hence 85dB SPL or above in the sweet spot is too loud for sustained critical listening aside from moving outside the ideal suggested listening curves. In a controlled studio environment, an SPL of 83dBC is more critical in the mastering context when dealing with compressed music most of the time. If your room is smaller than the norm, you may find the ideal SPL is a little lower at 78–82dBC. Make note of your experiments when finding the ideal listening level by observing the transparency of your audio references in your environment as suggested in the previous chapters. This should act as a guide to the SPL used in calibration. But you have got to start somewhere, and working at an SPL of 83dBC initially if you're unfamiliar with the concept will help you grasp the correct area for listening, give or take a few dB for your environmental circumstances.

With testing and critical evaluation, you will find your ideal. It is also important to note that when measuring each speaker independently, half the volume (turning one speaker off) would measure on average SPL around 77 or 78dBC because it is reduced by half, but the environment diffusion or reflectiveness plays on this measurements technical ideal. But our impression of loudness will not be half as our subjective impression of loudness will be more like -10dB as the SPL is half. This again is not the same as voltage where halving would reduce by -6dB when playing only the left channel after previously playing stereo. Subjective loudness, amplitude, SPL and acoustic loudness are not the same but are inherently interrelated. If you are interested in this functionality in the principles of loudness, you need to look towards audio acoustic publications. Here I am trying to check the directives focused on the most important aspect for audio engineers – maintaining equal loudness in all observations by using a fixed listening level.

I don't think that there is a correct level for everyone, though nearly all of the time when asking people to find where the music sounds best, it falls into the 83dB in stereo playback in the sweet spot. You need to find your optimum critical listening level through experimentation with references, let the audio analysis do the work and find where it sounds best. You'll be listening at an SPL of 83dBC give or take a dB once measured.

Cutting past the maths of it all, the conflation of too many standards and methods of measurement means this can be a confusing area to start to engage with. The most important aspect is to become aware of your own interpretation of subjective loudness and utilisation of calibration to get hold of that ideal to start working effectively at fixed listening level and equal loudness all the time. People will argue about the 'correct' measure, but that 'correct' measure is your listening context. Applying the methods explained will enable you to find the best outcome for your listening context, hence you will critically be working at equal loudness. It all comes down to that, the critical aspect in our listening methodology.

The monitoring controller

Many professional monitor controllers have an analogue switching matrix to facilitate 'listen' or 'audition' functions to observe different sources or parts of the matrixed signal path. These also allow the user to switch between multiple monitor speakers or even switch in/out auxiliary subs. In recent years there have been some positive developments with these types of features being integrated in the DAW monitor control management systems. But personally I prefer to control in the analogue realm. This is to maintain maximum quality from the DAC and secondly to achieve seamless auditioning between sources in comparison. I will explain both these aspects in practical use later in the chapter.

One advantage of the step controller is that the offset trim control can be set to make our calibrated level default to read zero. For example, if the monitor control read -9dB at 83dB SPL, the offset should be set to +9dB, making the normal monitor control reading 0dB at 83dB SPL to 'make up' the scale +9dB. Making our decided metered dynamic range equate to 0dB on the controller helps in changing the dynamic range quickly and efficiently in calculating its relative level to maintain our ideal listening level, as an example, if the meter was calibrated to -23dBFS relative to 0dB on the monitor volume. In changing our meter level to -10dBFS our audio would need to be turned up +13dB in the DAW to match the new average RMS level on the meter inside the DAW. To maintain 83dB SPL listening level, the monitor control would have to turn down -13dB, thus maintaining listening level, but the dynamic range available in the DAW has now changed to 10dB range from 23dB range.

When investing in a monitor controller, make sure all the levels can be trimmed to facilitate equal loudness in comparative playback. This maintains equal loudness throughout any switching and processing comparison. Equally any monitor controller should be absolutely transparent, it is the portal for the sound and not a processor to be changing it. Monitor controller features make our life easier, but it is the knowledge and use in application that is important. If your budget is tight, a perfectly functional monitor matrix can be set up in any DAW with consideration. The Nugen Audio SigMod plugin can be a helpful tool to achieve parts of this. Though this volume controller is only one important component of the monitor path, the overall quality of the monitoring path signal is governed by:

Monitor DAC > Analogue volume control > Monitor speakers.

The volume could be negated from this list by using the internal digital level control, but then in adjusting the volume in the digital realm, it would reduce the signal level into the DAC. This means it cannot work as efficiently as it is closer to its noise floor. Ideally the DAC wants to work in the positive dynamic range with its incoming digital signal. This is then converted using the most range from the DAC, and an analogue volume control simply attenuates the signal (no amplification means less noise/colouration of the signal). This gives the cleanest path to the monitor speakers and gets the best out of the DAC. But if the budget is tight, using the DAW stream as the monitor volume is a simple and clean solution.

The more control achieved over our monitoring scenario, the easier it is to maintain good practice and adapt workflow depending on the job at hand. For me, the features I need to work effectively and efficiently are:

- The ability to work in dB with the monitor volume, whether in a digital or analogue scenario to enable quick and accurate calibration of level to change the available dynamic range in the system; all while maintaining constant listening level.

- A facility to meter outputs accurately with multiple scaling options.
- Matrixing the audio between Stereo, Left and Right, Mono and Difference while maintaining equal loudness. Without these observations, it is very hard to master any piece of music as the analysis inherently lacks this perspective.

Considering this final point in more detail, mono and difference are simple summing matrices, though to maintain listening level there are implications that require investigation.

Left summed with right is really just two mono channels added to another mono. Whenever adding audio together, it will inherently increase the amount of energy in the final sum. If two mono signals at -3dBFS peak are summed to one channel, the outcome would be 0dBFS because it is twice as loud in the single channel. Hence, if two signals at 0dBFS (a normal audio reference) are summed to mono, the outcome in principle would be +6dBFS, but obviously in a digital construct this would just heavily clip as the signal cannot go over 0dBFS. This clipping does not happen when you hit a 'mono' button on your DAW or monitor controller or master bus. Thus it must reduce the level by -3dBFS on each channel before sum to achieve an appropriate outcome without clipping.

In terms of playback, this mono outcome needs to be distributed back to both speakers, thus maintaining the perceived level back to the original as the audio channels have been doubled the same as the original playback in stereo. This means when pressing the mono switch, it sounds about the same level as stereo. But if you compare a few differing mono buttons in your available paths, DAW master, sound card master, monitor controller, you may notice they are not all the same level when in mono sum. – one is louder than another. This is because when summing to mono, energy is lost to what would be in the difference. Remember when in creating a difference sum, the left is taken away from the right, meaning all the mono sounds cancel against each other as they are in opposite polarity. The remaining audio is all the aspects panned; the harder they are spread the more they will appear in the difference. For example, given a rock mix, the guitars are likely to be double tracked and hard panned, all the main vocals, bass, and kick and snare are likely to be down the middle and cymbals (the metalware), toms will be spread across the stereo image along with the general effects

and ambience. In observing ch1/L it would have half all the mono information plus one guitar and the L hand aspects of the overhead and effects. The
opposite is true for ch2/R, meaning in making mono the following sum and
gain structure will be created.

Ch1/L	Ch2/R	=	Mono Sum	or	Diff Sum
Guitar L		=	1 x Guitar L		1 x Guitar L
	Guitar R	=	1 x Guitar R		1 x Guitar R
Kick	Kick	=	2 x Kick		0 x Kick
Snare	Snare	=	2 x Snare		0 x Snare
Bass	Bass	=	2 x Bass		0 x Bass
Vocal	Vocal	=	2 x Vocal		0 x Vocal
Cyms L		=	1 x Cym L		1 x Cym L
	Cym R	=	1 x Cym R		1 x Cym R
Effects L		=	1 x Effect L		1 x Effect L
	Effects R	=	1 x Effect R		1 x Effect R

Clearly from this simple addition chart, it can be seen that in mono, anything hard panned is half the volume and that is why the guitars go quiet
when mono is pressed in observing this downmix to mono. In effect whenever mono is made from stereo, all the hard panned elements are losing 3dB
gain relative to the mono aspect where there is twice the signal. In listening
to the diff, there is no vocal, kick, snare or bass as they are cancelled out,
but there is still one of everything hard panned. Therefore, adding this mono
and diff back together (Mid/Side decoder), the stereo mix balance would be
returned as nothing was lost, the audio was just matrixed in a different perspective. It is important to understand this fully in using sum/difference and
mid/side during processing. We never lose any signal, it is just presented in
a different balance in the matrixed Mid and Side.

Coming back to our 'mono' button, we are working in stereo and listening
to a well panned mix with a positive stereo impression. By pressing mono,
the sum will sound quieter because of the energy lost to the difference,
meaning equal loudness is no longer being maintained in our mono matrix,
leaving the impression that the mix sounds dull or lacks lustre in comparison
to the stereo. To compensate for this, it is not unusual to change the relative sum power from -6dB overall to -4.5dB (which I personally prefer). But

this setting is normally user defined on a quality monitor controller system. The important aspect is to achieve equal loudness perspective in the mono impression that sounds correct to you. In digital, the Nugen Audio Sig Mod is an excellent tool to facilitate this to create a custom DAW monitor matrix.

The difference is a simpler outcome as it will always have less power than mono, generally lacking bass, this would be expected because low frequency will be in the mono element as a theme. But there are differences in how this can be matrixed to the two speakers, either maintaining the polarity difference, i.e., Left = Diff -, Right = Diff+. Personally I do not like this listening perspective, I prefer not to observe a phase difference by choice. The better outcome is to polarity switch the Left to achieve accurate observation of the S in the same phase out of both speakers. Hence as long as this difference is played out of both speakers, it will maintain a positive in comparative playback because it sounds very different to the mono aspect. This is unlike the stereo/mono comparison where the musical perspective should not be lost. It is important for any master to maintain the best mono compatibility possible.

The L/R matrices are uncomplicated, just in observing left, is simply playing that signal out of both speakers, thus maintaining listening level – the same goes for the right. Switching between these can often offer up additional observations about tonal differences between hard panned elements in a mix, such as guitars or synths, which may have bias in tone not initially obvious in the stereo or sum/difference.

In comprehending this matrixing, making these outcomes in the DAW monitor section or creating a bus matrix in the DAW as our default setup to assist with monitoring is straightforward. To achieve the required observations of the audio, it is the knowledge in approach that is important, and not the price tag of your monitor controller. Again the gear makes it easier, but it is not necessary to achieve the same outcome in critical analysis.

However you decide to go about controlling your monitoring path, when the signal is playing out the speakers in the air around us, it will also importantly be affected by the environment.

Environment

As a theme, a lot of mastering studios use free standing monitoring rather than soffit mounted (recessed) like in many modern post-production and mixing/recording studios. One rationale for this is the ability to shape the environment to our needs. More so I feel it just sounds better. The sound develops around the speakers in the room rather than being directed outward and absorbed behind the cabinet surround. Both these factors affect our perception of the aesthetic of the audio considerably. People will argue the semantics of this forever, you just need to test and listen. The decision is yours to make based on what you hear and not what someone calculates as an ideal or tells you is correct. You are doing the listening, and in the end you know what sounds best to you. As long as you are being critical and observing good practice around loudness, your evaluation is correct.

The sound level emitted by monitors is a critical part of the audio chain as discussed, often there is a requirement to 'hear' aspects of the audio that have been overlooked. One part of achieving this is the noise floor. Mastering rooms are very, very quiet, much more so than any mixing room. No equipment fans or noisy power supplies should be present in the listening space to cause issues. A closely located machine room for the equipment is very helpful to achieve this.

A well designed room is at the core of any good listening space. You do not fix a space, you make the space, and the treatment helps to create the overall vibe to the sound. There is not a correct answer, but straightforward principles apply – the smaller the room, the harder the bass modes are to design out. If the space is very large, a high level of sound absorption is required to reduce the reverb time. There is not an easy calculation to give the correct reverb time, most listeners just want an even balance across all frequencies that does not feel unnaturally dry. This is actually very hard to achieve. Equally the modal characteristic of the room needs to be evenly distributed to affect an even feel to the environment. The worse the basic room design, the more grouped or resonant the modes become. Acoustic and sound control is a whole science in its own right. If you are interested in this area, there are many publications on acoustics to get you started on that journey. The environmental considerations are critical to our monitor path, as is the studio monitoring.

Monitoring

The mastering room monitoring choice is a big consideration. From my own perspective developing as an engineer for decades, monitoring can sometimes seem to be an overlooked aspect in some recording studios.

I have been to studios where I wanted to adjust the monitoring setup to achieve a more critical appraisal of the audio. To improve any mix produced for delivery to mastering, correct the adjustments to set the stereo field and to align the speakers to achieve the right pitch to the ear. With less to correct, the mastering engineer can spend more time on the overview of the sound rather than fixing errors in the mix caused by a poor monitoring setup.

One aspect to consider is 'all monitor speakers are not the same'. This is an obvious statement but one that requires your attention. Speakers reveal the sound. Until you have listened to a high-end speaker system in a controlled environment, you have never heard the possible depth in music. It really is a revelation and one you must experience to start to fully appreciate the content of this book to its fullest depth. I still remember the wonderment experienced during my first attended mastering session in my formative years as an audio engineer/producer. I had never realised how different the clarity in the sound reproduction could be, and the detail in the mid-range was just sublime.

The nature of engagement in any field of art is in part the intention to progress. If you seriously want to develop your audio knowledge, you need to invest time in getting the most out of your monitoring path. Time spent listening, adjusting and evaluating the differences will pay dividends in the long run.

Take a look at mastering studios across the world, and a monitoring selection theme will appear. ATC, B&W and PMC will be popular names you will come across often as well as some bespoke setups. But it is no secret why these manufacturers are prevalent. They all deliver in phase precision in the sweet spot and an accuracy in the frequency balance. This is particularly true in the upper mid-range detail that means the engineer can 'hear' the true nature of the source material. This is equally obvious in the transient detail as the driver response times are quicker than lesser monitors. These monitor systems come with a hefty price tag, but once the investment is made, they can last almost a lifetime. In developing your studio setup, it is important to note the difference between monitors by listening, the environment has just as much a factor to play as the monitors. Without adequate control of the room acoustics, a three-way monitor will not reveal significantly more than their two-way equivalents. All these aspects need to be considered together in any budgeting to achieve the most effective outcome possible.

This leads us to the final aspect of our monitor path – monitor DAC.

Monitor DAC

To achieve an effective listening environment, the consideration of the conversion between domains must not be overlooked or assumed. All converters are not the same. Another obvious statement, but the better our monitors and environment the more the quality of the DAC becomes exposed, making the differences clearer. In comparing a normal multi-channel DAC on a quality monitoring system against a dedicated stereo DAC, the first thing that will become apparent is that the stereo imaging is improved because it is simply just in time across the frequency ranges, unmasking the sound. There is not a correct or best DAC, but they all impart some different aesthetic on the conversion process. Leading manufacturers such as Prism Sound, Cranesong, Burl, Benchmark, Lavry and Lynx Systems all make excellent converters but they do not sound the same. As with speakers, the sound an engineer prefers comes down to their critical evaluations overtime and comparisons made.

Most professional DAC generate their own clock internally. They sense the incoming digital stream sample rate and bit depth. In reading the stream into a buffer, they use their own clock to resample the playback to the converter accurately. This rejects any interface jitter that may be present in the system as the signal has been 'retimed' to the correct sample rate using the DAC's own clock. You can read more about jitter, its effects and system clocking in Chapter 10 'The transfer path'.

All this means there is often no word clock input option on a dedicated monitor DAC, unless it is a combination ADC/DAC which will likely have these clocking options. It is important to make this distinction at this point. It is helpful to have separate systems for each DAC and ADC as to decide on clock and resolution independently. I will expand on this concept in Chapter 10 'The transfer path'. There are also clear advantages to being able to select between multiple paths with the same type of DACs. When the DAW's DSP is loaded and playing, it means it is possible to make seamless A/B between the transfer path and original audio source because the DSP is not loading and unloading as will happen when switching with solo in the DAW. But it is critical the DACs are identical to maintain integrity in the monitor path comparison.

Summary

There is a straightforward thought at this juncture in our discussion. If you can hear it, you can fix it. If you cannot, you will not know it was there to be fixed. This simple observation is one of the fundamentals in starting to effectively master audio. This does not mean you need to spend tens of thousands on speakers or monitor controllers to finalise audio, but clearly there is a need to carefully consider the monitor path and environment in any budgeting.

It is also worth noting room correction software at this point. These systems use EQ and phase to 'correct' the sweet spot for an ideal listening outcome in principle. The difficulty is that our head moves, even a millimetre difference changes our perception of phase difference when monitoring. Hence these systems on the face of it can seem to improve the sound, but having tested several systems I would opt to change the environment and monitor position rather than put an EQ in the way of the sound. This is not to say these systems cannot be useful, but if you want to really become critical in your listening you need to focus on improving the monitoring/environment/DAC and not putting something else in the way. The more effective our monitor path and environment, the better the potential outcome in analysis and the proficient application of tools, yielding a better sonic outcome overall.

If you really want to use this chapter's information, you have to get stuck in practically. Listen, measure, test, reflect and evaluate. Keep this up with quality sources from different genres and you will fully comprehend calibration and be able to use your monitoring in the most effective way possible. In doing so, become self-aware in how to improve your outcomes and most importantly gain real confidence in your own listening ability. This will intrinsically translate to the audio you created. Remember all the knowledge discussed is a guide, trust your listening above anything else. Only you know the correct answer in that regard.

Looking to the mix

There are some elementary principles to receiving and requesting outcomes from the mix engineer. The first is about resolution. Today and in the future, it is unlikely you are going to come across a scenario where the source of a new project is not a digital file format (unless you are remastering from historical sources). Either way, it is crucial to fully comprehend the importance of resolution.

Resolution

In this modern digital construct, most mix projects are completed in a DAW and likely to be running at 64 bit in the mix bus. Even if the mix is external in analogue, or a combination of analogue inserts and digital pathing, the audio will eventually come back to a DAW via its analogue to digital convertors (ADC). This means the higher the mix export resolution from the DAW, the more of the original mix quality there is to master with. But equally it is important to comprehend the source system. A DAW project will be playing at the fixed sample rate selected in the initial creation of a project. This is likely to be set based on the interfacing and available digital signal processing (DSP). Remember, doubling the sample rate doubles the DSP load, i.e., halves the number of potential plugins and track count available.

In this regard, it is best to request the mix engineer supply the mix at the same sample rate as the project setup and not 'upsample' from the rate they have been working at during the project's construction. This will avoid potential errors as some DAW project plugins do not perform correctly on being adjusted in rate and will potentially deliver incorrect

DOI: 10.4324/9781003329251-5

outcomes without reloading as a new insert. This is aside from the computer resources being at least halved in upsampling leading to DSP overloads and disk read errors as resourcing is low. This is different to upsampling a file that is already rendered, as another set of conditions apply and the quality and method of the conversion used are crucial. Hence this is best left as a decision during the mastering process and not the mix. Equally rendering at a higher samplerate in the bounce setting is not going to capture anymore data, it is just resampled at the output, which again could cause errors. I will cover these parameters in later chapters. For most output formats now post mastering aside compact disc (CD) audio, the delivery will be at 48kHz or 96kHz, making it better for the mix engineer to work at these scalable rates at the start of a project where possible. These formats scale to video media and HD audio ingestion without conversion. It is better to scale down to CD quality from the outcome with the highest quality sample rate conversion software. Some mastering engineers prefer to do this in the analogue transfer path, sending at one rate and recording back in on a separate system with its own independent clock. This scaling should not take place in the DAW mix bounce. If the engineer has worked at 44.1kHz, it is best to stick with that but I suggest they think about more appropriate rates at the start of future projects.

Higher rates are possible at 192kHz or faster looking towards Digital eXtreme Definition (DXD) processing systems such as Pyramix. The best quality clocks are required to run at these rates and sometimes dedicated hardware. But just because a piece of equipment says it can do something, does not mean it can do it well all the time. Rigorous testing is needed to be sure there are no negatives to the intended positive of a higher resolution, which in principle is always a good thing. There have been many transitions from lower resolution systems to higher in the past, just in my time working as an audio engineer. Testing and developments will eventually make these transitions happen again in the future.

In terms of bit depth, the higher the better. The bigger the bit depth, the more of the original master bus of the DAW will be captured. As most are now working at 64 bit, 32 bit floating point is ideal as a standard for export in this case. This formatting means the capture will contain all of the mix bus and the file will also retain any over shoot above 0dBFS. Not to understate this, 24 bit at 6dB per bit gives a possible dynamic range of 144dB, 16 bit 96dB, but 32 bit float has 1528dB of dynamic range, of which over 700dB is above 0dBFS. It is almost impossible to clip when bouncing the mix for mastering. If a 32 bit file looks clipped, a normalise function will readjust the peak back to 0dBFS and you will see there is no clipping present in the file.

If the source is digital but the mix is analogue, the converter used in the transitions between domains governs the capture bit depth. For the majority of commercial ADC, this will be 24 bit, which is a 144dB dynamic range. It is more than we can hear, but care should be taken to use all the available bits. If aiming for a peak between -3dBFS and -6dBFS, all the possible dynamic range will be recorded while also avoiding potential clipping at 0dBFS. Some meters are not the best at showing clip level and some converters start to soft clip at -3dBFS if it has a limit function. Remember recording at a lower level and normalising in the DAW does not put back in what was never captured, and recording at 32 bit with a 24 bit converter is still only 24 bits in a 32 bit carrier. The DAW bit depth does not govern the captured path, the ADC does. But changing the recorded file with processing is when the higher bit depth comes into play. Even in a simple reduction of volume, it means all the original bits are still intact as there is dynamic range left to turn down into. If the DAW mix bus resolution was only 24 bits, the amount the audio was reduced at peak would be cut off at the bottom of the resolution. This is called truncation and should be avoided. In a modern DAW context, the mix bus is working at 64 bit, meaning a recorded 24 bit audio file could be reduced by 144dB, rendered at 32 bit float in an internal bounce and turned back up 144dB, and there would be no loss in quality from the original file as the headroom is easily available. To avoid losing any of this quality gained from the mix, it should be exported at 32 bit float. The way to deal with bit depth reduction to formats such as CD audio is discussed in Chapter 16, 'Restrictions of delivery formats'.

An aspect worth exploring at this point regarding resolution and truncation is offline normalisation. If you are recording a file into a normal digital system from analogue, the converter will be working 24 bit, if the recorded level peak was -12dBFS and each bit is 6dB. The file is actually 22 bit, recorded at -24dBFS peak and it is 20 bit, -48dBFS peak 16 bit and so on. The capture level is critical to the quality. This simply cannot be put back in by normalising the files to 0dBFS because it was not recorded – there is nothing there. In applying normalisation to the file, the bottom of the floor is raised and in doing so quantisation distortion is created between what was its floor and the new floor. With the 16 bit captured example, this is a significant amount and clearly audible. This adds a brittle sound to the audio in the same way truncation does when the bit depth is reduced. This is why normalising offline is the wrong approach when trying to maintain quality. It can be very interesting creatively in sound design to destroy or digitise a source in a similar way to a bit crusher, but this is hardly the sound that would be wanted on any part of our master. But if using the fader to turn up the audio or the object-based volume to apply an increase, our DAW is working at a much higher resolution of the mix bus, and this quantisation distortion is minimised.

Whilst discussing resolution, it is important to mention Direct Stream Digital (DSD), which is another method of encoding from analogue to digital. DSD is a one bit system, in a simple sense it means each sample stored can only have two states, which are read as go up or go down. With a super fast sample rate of 2.8224 MHz, 64 times the rate of CD audio, each time a new sample is taken it just goes up or down in amplitude relative to the next. In many ways it can just be thought of as a redrawing of analogue in the digital domain. But the super fast sample rate means the recording cannot be processed by a normal DSP and requires conversion to DXD at 24 or 32 bit. The lower sample rate of 8 times CD audio is at 352.8kHz. Higher sample rates come with their own potential problems, which I discuss in Chapter 10 'The transfer path' and I discuss the practicalities of DSD and DXD use in Chapter 16 'Restrictions of delivery formats'.

Source dynamics

The more overly compressed the source, the less range there is to set the trigger level for any dynamics control. As an example, the mix engineer might make a 'loud' version for the artist to listen to the mix in context with other material, because louder sounds better, but equally, the artist will be influenced by the loudness difference between their mix and commercial audio in their playlist. Hence this 'loud' version works well in helping the artist understand the context and signing off the sound balance in this more realistic comparable context. But this is not the source required for mastering, though in experience, it can be helpful for our context to have a copy, because sometimes our ideal master is less pushed in level than this 'limited' version. If everyone has been listening to this limited 'loud' mix and our master is delivered quieter, without a conversation to explain why an incorrect comparison can be made, louder will sound better in principle. Primarily from a mastering point of view, if our source is an 'unlimited' version, the decision to apply a peak reduction at the start of the chain is now our choice rather than it being dictated upon us. This is a good outcome; remember that dynamics work from a threshold, and the more its range is restricted, the harder it is to target the required aspect.

With less experienced mix engineers, they may be applying master bus processing to make the mix sound more like the outcome they want. Not a bad thing in principle, but the point of a mix is you have access to the individual tracks, if something is not compressed enough, it should be applied to the individual sound or the group bus, i.e., a drum or vocal bus. It is ill-considered to apply it to the master bus. It can seem effective on face value, but this is just skirting over the issues within the mix. If they have applied a multiband and it is triggering in the bass end with a wide variance in gain reduction, they should instead be targeting this element in the individual channels where the sub/bass is present. Unmasking the rest of the mix which was over processed by the master bus multiband. This is apart from the more effective trigger achieved targeting the original source tracks alongside better evaluation of the envelope required.

Sometimes a mix is so reliant on the master bus processing it is pointless to request a mix without, because the majority of the sound balance is happening on the master bus. In this case, you just have to go with what you have. By undoing the master bus processing, the mix will no longer be coherent, especially in context to the sound the artist and label have signed off on. There is a big difference between skimming a few peaks off and applying a multiband compressor to your mix bus.

This should not stop mix engineers from using a bus compressor on their master bus if they want to 'push' their mix into it. But from experience, a lot of engineers get out of this habit or deliver a 'clean' mix alongside without any process on the master bus. The mastering engineer in principle should have a better compressor or similar to the same one used, but with a different perspective to achieve the 'glue' being sought by this type of mix bus compression.

Top and tail

If a mix is submitted with a fade in and/or fade out, the relative audio level is not constant during those fade transitions. Examining this from a processing point of view, if the audio fades out, it will fall through the threshold of any dynamic control. If this is a downward process, the likely outcome is a period of hold that will be created in the fade. The material above the sound will be reduced in gain, when it falls below the threshold, the compression stops, meaning the fade fall level is reduced until the gain reduction amount is past, and the fade continues. The fade will now have a hold period inserted in the fade due to the compression reducing the amplitude then not reducing past the threshold. The same principle can be applied to static effects such as EQ, as the amount of compound EQ will change relative to the transition level during the fade.

To avoid this, it is best to request the client send a version with and without fades. A master can be made and the fade applied mirroring the client's original musical intent. This sounds more effective on long reprise fade types, as the whole sound of mastering is maintained while the fade out transitions. This is in contrast to processing with the fade in place, as the impact of the mastering change gets less as the fade goes through transition. Just as a straightforward observation, the fade would transition through the threshold of the dynamic tool, meaning the processing timing imposed would change. The start of fade dynamics would be correct, but by the tail end there is no processing. This is not the best requirement for a fade where the RMS content is constant. Equally if there was an expander enhancing the width in the S as the fade reduced, the track would become narrower.

In every regard, it is better to have a mixed version without any main fades; these are always better applied post processing. This does not apply to break down ends where the instruments are turning off in turn as part of the music's closure. But it applies when the mix engineer is fading the master bus to achieve a fade or applies a post fade to the file to give the client the outcome they intended musically. This is fine for the artist's 'limited' reference, but the fadeless version is best for the mastering chain. It can be faded to match the artist reference afterwards.

It is also worth noting, if the mix engineer can leave an audio header at the start and end of the bounce, they will have captured any possible broadband noise present in the system. This can be used for noise prints if any reduction noise is required, but mainly it means the top and tail of the actual audio will not be cropped when supplying the mix. Remember when mastering the level relative to the noise floor could increase by over 20dB. Hence the tails of the audio, which sounded fine, can be cut off as overly zealous editing has been committed to the mix file. The end of the tail will be stepped, but if a lead in/out 'blank' is there, and the file is 32 bit float, the audio's natural fade will be retained below the original audible spectrum.

This potential pull-up is also where noise can come from especially in the tail of a fade. Sometimes a little noise reduction in the final transition goes a long way to smoothing this outcome post mastering while creating the fade. Noise prints in this regard are not always required as a dynamic frequency dependant expander like Waves X-Noise, or an advanced algorithmic tool like Cedar Audio Auto Dehiss will work very well. This is clearly the best broadband noise reduction I have ever found. For more intrusive sounds such as static or broadband electrical noise (guitar amp buzz), the TC Electronic System 6000 Backdrop noise print algorithm can be very effective. Equally all the Cedar Tools are top of their class in any form of removal of unwanted sonic material.

Parts

It is best to request instrumental mix masters alongside the main mixes. One of the core uses for music in the modern era is its role in conjunction with video. This could be a television show, film or advertisement. All this falls into a category called synchronisation of music (sync). This is often directed or managed by a music supervisor on behalf of the media/film production team. Making sure instrumentals have been considered can save the client hassle, time and money down the line, but obviously the creation of extra mix versions is an additional use of time that the mix engineer should be charging for.

The master supplied into this post-production scenario is no different from the other outcomes considered. The music would be 'mixed' in 'post' with any dialogue, sound effects and so on, often this would be at a much wider dynamic range outcome than the music itself. Especially in a film context, there is no rationale for this to differ from the music the consumer would digest, the music supervisors have selected it because it sounds good, mastered! However, it may require editing to better fit to the visual or just have the music without any vocals in sections. An instrumental master is vital to be able to remove the vocals from this post scenario where the producer will want to create differing sequenced events to the original. It is very helpful if these two masters are sample aligned to make that edit process seamless for the post engineers even if this means the 'top' of the file has 'digital black' where the vocal intro may have been. Digital black means there is no signal in that part of the audio file. This term can also be used where video has no content.

Having a sample accurate instrumental master also makes for the easy creation of 'clean' versions where any potentially offending lyrical content can be easily removed without fuss and a new 'clean master' rendered post mastering. The artist may want to supply clean versions edited with new lyrics or effects in the gaps. Though I found over the years for the many differing radio/tv outcomes, context is key. It is usually highly dependent on the target audience and demographic. For example, I have had to edit out the words 'drunk' and 'damn' before for radio play. Equally, editing down of the solos breaks for radio is not uncommon and having the instrumental without any vocal overlaps has made this easier to edit as the music transitions between sections. You could say 'why not go back to the mix' – cost first, time second, and third recalling all these differing paths from mix to mastering means some of it might not recall as the original makes it sound different. Simply put, just have an instrumental. It saves a lot of time and hassle in any scenario.

The mastering engineer's remit in my view is to have the foresight and awareness of potential issues for a given project. It will put you as an engineer in good standing with your clients being on top of any potential future scenarios that may come to pass. Even in the simplest sense of making the client aware, they may need instrumental mixes, even if these are not processed at the time of mastering. It is a lot easier to recall a mastering transfer path (something done all the time) rather than trying to pull a mix project back months later on a different version of the DAW with older plugins outdated or unavailable in an updated studio where access time can be limited.

The final aspect where an instrumental version can be useful in the mastering process is the simple separation of the vocals. A straightforward sum/difference matrix will polarity cancel the vocal out of the main mix. This render summed with the instrument is the same as the main mix, but take care to make sure the polarity remains true to the original. Now any targeted processing required on the vocal or the mix can be applied without affecting the other. I find this can be helpful, though in truth, I would prefer to work with the mix engineer on revisions to get the mix right from the start rather than being this intrusive to the mix balance. The next logical step in this approach to changing the mix balance would be to ask for stems.

Stems

Stems are a breakdown of the mix or master into its composite parts. This term is used to describe the final 5.1, 7.1, 10.2 or Atmos delivery for audio use in sync with video. It could also mean the stereo parts of drums, bass, instruments and vocals of a final stereo mix. The important part of the latter is that it should, when summed, add up to the original mix. In a sum/difference with the original stereo mix, the stems should null. This can be quite hard to achieve if sidechaining is taking place across the mix parts and some plugins have a random modulation which does not replicate each time and so on. I can see a clear point in mastering these 'parts' if the intent is to use them as such in post-production as with the instrumental. Take care not to confuse this with what is often called 'stem mastering'. For me, the mix is something that is complete, all parties involved have signed off as the sound balance required. The mastering process enhances this 'good' mix to make it 'fit for purpose' in delivery – mastered. In being asked to change the mix, I am no longer mastering, but finishing the mixing process and then putting this into a mastering chain albeit engaged in these two processes as the part of one whole chain. This mudding of the waters can lead to an ambiguity about the role an engineer has been hired to perform, hence expectations of outcome, cost versus time to complete and revisions to apply. It is no longer just the master processing that might be asked to be revised, but the whole mix in any aspect.

A client might say they are delivering a vocal stem and backing stem because they just want you to balance the vocal into the backing when mastering. The vocal is an integral part of a mix and the mixing process, not something to be simply sat on top with no ability to balance the other parts around it, especially dynamically. This naivety should not be dismissed. Everybody has to start somewhere, and knowledge is often gained by outcomes being challenged. You should support the client and direct them towards an appropriate mix engineer to address their sound as a whole. Then it may come back to you for mastering.

In my experience it is better to work with the client to achieve the mix that everyone wants to sign off on, and then master it. This is a clear demarcation between mixing, achieving the correct sound balance and then mastering, making sure it translates well on all systems. In entering into stem mastering, these lines are blurred. How many stems are too many? Surely to tweak the mix fully, every sound should be printed out with its effects and processing embedded to be able to 'tweak' the balance. But logically, the next step could be to have dry and wet stems, as a better outcome could be achieved again. Before you know it, you are in a full mix process with the original multitrack master post recording. This is fine, as long as you are very clear with the client where this is going, and it is going to take a lot more time and hence increase cost significantly. Realistically you are just mixing the song in its entirety. More importantly the overview brought by the mastering engineer's perspective can be easily lost in the details of the mix. Why did the client come to a mastering engineer in the first place? To gain a different perspective in overview, not to lose it!

Personally, I try to avoid these scenarios ever developing by making the roles clear. If it is a requirement, the client will know the cost ramifications. I also feel the mix should be complete. If you as an engineer/producer are finding you have to deliver stems to get the audio to sound right in the end. You really need to start looking for a mix engineer to help achieve that outcome, getting the mix ready for mastering by the mastering engineer in my opinion. Maybe you are a great songwriter or programmer who has brilliant production ideas, but this does not mean your mixing (sound balancing) is up to the job. Why do you go to a mastering engineer? Because they have a skill set. The same is true of the mix engineer. In creating art, we should seek people who can assist in taking ours to the next level. Embrace it, do not think you are losing control or giving something away in seeking help.

Notes on instrumentals prints or mix re-prints

When discussing rendering outcomes with the mix engineer/client, it is important to be clear when a re-render is requested, it is in all regards identical to the original delivered other than the requested change. If there is even 0.1dB of overall loudness difference, it affects every part of the mastering transfer path down the line, meaning the outcome will not be the same. This is extremely unhelpful. Over the years, I have developed the habit of sum/difference testing of any revised mixes in a separate project back to the original mix to confirm their validity in gain structure before use in the final transfer path. This applies just the same with any instrumentals, cleans or sync versions supplied with the original main mixes. Much better to spot these mistakes at the beginning of a project than down the line where time constraints are normally tighter and deadlines are to be met.

Signing off

In working effectively with a client, it is important to make sure sources are valid and the client understands that the mixes once supplied are final. To make changes involves a cost. Whether this is a minor tweak to the mix or a major change, it has a cost implication. It all takes time and unfortunately time is not free! Making this crystal clear at the start avoids any ambiguity further down the road. In the era when it would seem that a recall and tweak can be made at a whim because of digital technology, the assumption can be that reprinting a master is a straightforward process. But mastering involves listening, and this 'real-time' aspect cannot be overlooked. Equally most analogue hardware does not recall itself. Making sure the client is aware, means they are clearly focused on delivering mix sources that are completely 'signed off' by all parties involved, thereby ensuring a smooth transition to the mastering outcomes. Our outcomes need to be signed off before making up all the parts (see Chapter 16 'The restriction of delivery formats').

Notes on remastering sources

There have been many different delivery formats and associated resolutions over the decades, both in supply to mastering, actual master output and the final consumer product for consumption. Most of the mastering engineers' output would have been a pre master format. This is a finished master before the final master created during manufacturing. For example, source mix master > pre master > glass master > physical CD. The term pre master is sometimes used in modern delivery to describe the mix master. Irrespective of the terminology, what I have learnt about how information is presented is absorbing the knowledge of those engineers who were involved in delivering formats at the time. This helps to understand the nitty gritty of the entire process, comprehending the potential issues involved in transfer and maintaining integrity of these sources if there is an intent to use them as a future source. This core knowledge is diminished with each passing year as the formats and the engineers involved in their creation become fewer and less vocal. To engage in breathing new life into the music of old, comprehending the purpose of the source in the production chain and its playback mechanism is needed to start to use these sources as new potential outcomes.

All this falls into a different trajectory for mastering – first transfer, second restoration and third what is called remastering. This is not to be confused with mastering new music as the aims and objectives are quite different. I will discuss the remastering and restorations approach in a future publication.

Mix submission requirements

In making a mix formatting request, I would ask the client to supply:

- All audio files at lossless 32 bit and same sample rate as the original project (24 bit if not possible). Use interleaved or non-interleaved .wav or .aif files for maximum compatibility. DSD or DXD format source is fine.
- Aim for max peak between -3 and -6dBFS where possible.
- No mastering bus processing or limiting for perceived loudness. Supply an additional master bus processed version if this is vital to the mix sound balance overall.
- Supply the client a 'limited' version if this is different to the main mix master.
- Leave a minimum of a 3 second pre post gap on the file. Do not hard-edit or top/tail.
- Do not apply fade in/outs, supply a separate faded version which can be 'fade matched' after mastering.
- Main mix and instrument versions or any other variants have been signed off by client/artist/label.

Mix/mastering feedback loop

As a mix engineer, the more you work with the same mastering engineer, the better they will be able to help you improve aspects of delivery and sonic balance to enhance your outcomes overall. One of the best things you can do is to stick with them for a considered period, at least a couple of years. If a mix engineer chops and changes between outcomes, they miss out on developing a rapport with a mastering engineer. This will help advance their mix technique and outcomes as they are able to have the direct and frank conversations required to improve with the context of a set of jobs.

Remember a mastering engineer is skilled at observing overview, and this applies to a mix as well as a set of differing projects. Themes become clearly apparent to us.

Personally, from a mastering engineer's perspective, I find these relationships very constructive for both parties, I understand their artistic vision more effectively, and they gain constructive criticism during development of their mixes – a better outcome for the product overall.

Analysis
Constructive, comparative and corrective

Dissection of a mix is essential in our ability to target issues effectively with processing. If primarily dealing with stereo sources, just listening to stereo is not delivering an in-depth evaluation. Afterall, a stereo mix is a composite of two parts – Left (L) and Right (R) – both of which in analysis contain different aspects of the mix, otherwise channel 1/L and channel 2/R would be identical, hence mono. The difference between elements in channel 1 and channel 2 is what generates our perception of stereo. In many ways, stereo is just the outcome, what the consumer digests. The actual processing path in mastering is two discrete channels, and it is helpful to start to think about it in this perspective.

The sum to mono (Mid) and sum/difference (Side) aspects deliver observational detail not always immediately clear in the analysis of stereo. Being able to zoom in to the composite parts of a mix makes apparent where instruments have been panned or placed, and makes clear what aspects are different between channel 1 and channel 2. This helps our observation of perceived depth of field and general placement of effects in the spatial and stereo impression.

Equally and most importantly, the phase coherence and mono compatibility of the mix's construction becomes clear. Downmix compatibility is a crucial element of a master and any form of post-production delivery. The better the translation of stereo to mono in mix construction is comprehended, the more the simplicity of multi-channel audio arrangement and downmix becomes understood. In fully comprehending matrixing stereo and mono, this knowledge becomes transferable to Ambisonics and Surround Sound systems such as 5.1, 7.1, 10.2, Atmos and so on. The same principles of phase relationship and amplitude apply.

DOI: 10.4324/9781003329251-6

Discrete channels analysis and link to application in process will become apparent as a critical driver behind the construction of any transfer path and the design of the majority of mastering equipment. To start to assess any mix it is critical to comprehend its matrixed parts from stereo to observe Left, Right, Mid and Side (LRMS) and stereo – overall what is referred to as constructive analysis.

Constructive analysis: Mid Side encoder decoder

It is useful at this stage to fully understand the relationship between Mid/Mono/M and Side/Diff/S. The creation of these matrix outcomes is crucial to analysis as much as it is when linking these observations to processing using the same paths.

'Mono' is a simple sum, L+R=M. This addition means the gain of the summed channels needs to be reduced pre to avoid clipping in the bus. For example, L-3dB+R-3dB=M. 'Side' is this addition but with the right channel polarity inverted, L-3dB+R-3dBØ=S. These two aspects together create a Mid Side Encoder (MSE), taking Stereo to MS, or L/R channels 1/2 to MS channels 1/2.

Now that the audio is matrixed in Mid Side, it is possible to process those aspects discreetly. To decode this outcome back to stereo after processing, the Mid and Side channel requires summing and reversing the phased aspect creating a Mid Side Decoder (MSD), for example, M+S=L and M+SØ=R. As the summing had previously been reduced to maintain headroom, this addition does not need a reduction in gain to compensate back to the original level.

Using this MSE and MSD in conjunction creates an MSED matrix path which can be observed in figure 5.1. As it is only polarity and amplitude that have been used to create and revert the signal, the outcome after the MSED should be identical to the input unless an aspect is changed within the matrix. When changing the gain on the Side and when listening to the LR stereo output of the MSED, it will sound like the image is wider and narrower in changing the gain as the relative energy between the Mid and Side is being adjusted.

It is easy to test if an MSED pathing is correct with a simple sum/difference null test to check that the input is the same as the output.

There are varied tools available to assist in the creation of these M/S matrices in the DAW. Voxengo MSED is a free plugin which enables the use of MS in a normal stereo channel path. The Nugen SigMod also works well though with more flexibility. Some DAW's like Sequoia/Samplitude have these matrices built in to the channel bus. In analogue, dedicated hardware such as the SPL Gemini M/S or most mastering consoles have M/S matrices built in. In addition, they often have a width control which enables adjusting the gain structure between M/S paths but while maintaining equal loudness. These are quite effective to set the correct impression of width without loudness effecting our judgment. Whether you have an expensive mastering console or are using a free plugin, it is the knowledge about how M/S works and changes the impression of the audio that is important and not the cost of the method used.

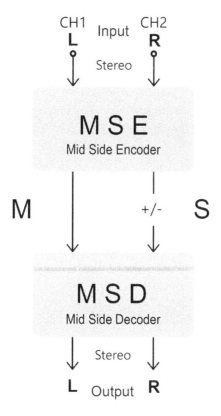

Figure 5.1 Mid Side Encoder/Decoder pathing

Constructive analysis: Comprehending stereo

In starting to fully understand the relationship between Left and Right, Mid and Side, Ch1 and Ch2, a phase scope can be a positive place to start. By observing a stereo mix playing through a scope, the majority of energy will be focused around the centre vertical line labelled mono. If the source is made mono before the scope, the graphic will only show a vertical line as there is no difference between Ch1 and Ch2 (L and R). In inverting the phase of one of the source channels, the energy is then displayed on the horizontal line, S+/S-. The audio is mono, but reversed in polarity 180 degrees. The relationships of these axes can be observed in figure 5.2.

When using a scope to visualise a mix, the more that energy is observed toward the 'S' line, the more negative phase difference the source contains. This is obviously not a good thing as the mix mono compatibility will be lower the more the energy there is on the horizontal 'S' line.

The scope also helps in the observation of L R context. The lines at 45 degrees are labelled L and R. When playing a stereo mix and muting the right channel, the signal will be observed on the L line no matter what the polarity is, as it is not summing or negating against another channel. Unmuting the right channel and muting the left causes the signal to flip to the R line. The swing between the L and R axis will enable observation of the bias in the amplitude weighting of the mix between channel 1 and 2.

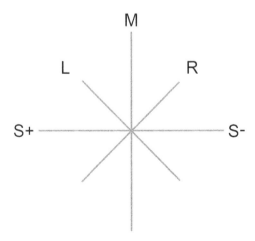

Figure 5.2 Phase scope axes of LRMS/Stereo

If the general RMS is louder on one channel than the other, a tilt will be observed away from the vertical mono axis. You can demonstrate this by playing a stereo mix and reducing the gain of either L or R channels 1 dB before the scope. A clear 'tilt' in the visualisation will be observed even with this small bias. I have found the 'Flux Stereo Tool' to be one of the best visual scopes to display the left/right stereo context.

In the days of analogue mixing, it was not unusual for there to be a slight energy imbalance between L and R because the mixer/monitoring or stereo recorder were not 100% balanced. Re-adjusting this 'tilt' would enable the correct focus in loudness to be balanced across both channels. Now with digital systems being the mainstay, these types of bias are normally observable and fixed in the mixing process. But, if remastering audio, this is something that must be checked especially if the source is vinyl, which will nearly always be bias in relative playback of L and R.

Constructive analysis: The line of treble

Refining the observation of the width of a reference in comparison to the mix under analysis is listening to the difference in stereo impression. In assessing, it should also be observed if the tone is duller or brighter in the stereo impression. For example, is it just the amplitude that changes in width, or a volume and/or tonal difference that is changing the width of the stereo field? With multiple references between the mono and difference, it will be observed that the amount of treble in the top end stays uniform when switching. But often if you make this comparison with a mix, the treble can be less in the Side or the opposite, with the mono being dull relative to the Side. I refer to this imbalance in treble across the stereo impression as 'the line of treble'. It means our mixes have an imbalance in the tonal impression between the mono and the difference, and the mix might be dull in tone overall or equally bright or bass heavy. It just means in analysis this imbalance is noticed in the context between the Mid and the Side. Hence when contextualising this to a process, it can be effectively fixed alongside any overall tonal change required from analysis. Often when an aspect can be focused out of the whole in analysis, being broken down in focus, it makes it easy to hear in the complete picture when it was not observed earlier. This is a benefit of using mono/difference during the analysis stage.

Why mono is still the queen and king

The consumer does not generally think about stereo playback in the way engineers consider the audio. People might buy a Bluetooth pod for playback from streaming, they may have a mini Hi-Fi with both speakers next to each other on a shelf in the living room, or one speaker might be in the kitchen and the other in the dining room. The list goes on of potential stereo failures in consumer playback, though these are only failures in our engineering view. I am sure they are very pleased to have the same music belting out in the dining room and kitchen when their friends are around. But there are also Hi-Fi consumers who will make a clear and considered effort to achieve a stereo impression. Though one place a lot of consumers will hear true stereo today is on headphones. But how many people actually use those for their primary listening? This is hard to say, but one thing is clear, there are a myriad of possibilities, as engineers, it is critical to be conscious of how our master will translate on all these outcomes in stereo and mono.

The better our mono compatibility, the better the outcome in these defused scenarios. Even the playback itself can be the enemy. If you are streaming audio, what is the easiest way to make the stereo signal have less data? Reduce playback quality? But if this is already reduced, why not just make a stereo signal with two channels a mono signal with one channel? Half the channels and the data requirement is halved. So even if you have a stereo system, it might be playing in mono in consumer delivery. The same applies in a car with a stereo receiver; this does not mean the amplification is stereo. An expensive car will have a positive playback system, but not everyone can afford an expensive car.

Using your ears, and not your eyes

Since the turn of the century, audio visual interfaces have become more intricate and informative, but this can often become a distraction and not a helpful tool in analysis. A whole frequency spectrum visualizer is not music, but rather just a visual representation of amplitude at a given frequency over time. This tells us nothing about the context of frequencies in relation to the music. Basically, it looks nice, but is not actually that helpful musically during analysis. In context, technology has not developed a system that can accurately measure relative musical loudness between songs, hence there is not a musical level meter, just a measure of RMS or peak at a given frequency. These visual tools are helpful as a guide when comprehending this premise, but they should not be used to make primary decisions about frequency balance in analysis. Our ears are much better attuned to give an informative musical impression with the correct listening and referencing approach. To achieve the listening level, an RMS meter is used, but to accurately level from there between tracks, our musical judgement is needed, instead of just looking at the visual or derived numbers. Using a full spectrum analyser during analysis is more of a hindrance than a help in my opinion.

Engineers of my generation and earlier had the advantage of training without these visual aids, except basic metering for maintaining appropriate signal to noise and headroom in the signal path. For those engineers brought up with new visual aids, I would suggest weaning yourself off their use as a mainstay and begin to really listen without looking in response to the audio. If you are unsure of whether you are or are not overly influenced by visual aids, look at your use of EQ. If you are always using a visualisation when EQ'ing, I would suggest turning it off when engaging in your next project. Use your ears and not the visual interface. If you cannot EQ without the visual, you are clearly not listening effectively. A self-imposed moratorium will in time deliver more musical decisions based on what you hear and want to correct or enhance, not what you can currently see could be changed. Remember, the visual is showing you amplitude over time and frequency so that your response is changing the volume of parts of the music. This is what dynamics are for, and not EQ.

Analysis requirements

When arranging audio files in the DAW to start to analyse and assess a mix. It is useful to set up a sum/difference matrix between the original mix and a copy to act as the source for the transfer path. A third track can be created for your audio reference material offset from the mix for easy access during playback. This avoids the possibility of reference and mix being played at the same time and potentially doubling the playback level. It is always best to work sequentially with different sources when mastering, rather than stacking tracks, as would be the focus when mixing.

Both the source and original track audio should be sample aligned to achieve null when both tracks are unmuted. This will be useful when starting the process because the difference can be accessed to 'hear' dynamics and equalisations effect on the transfer path. It also allows for toggling the solo on the original track when in mute using solo exclusive mode to compare to any processing directly on the source. This is a helpful comparative to maintain equal loudness during processing.

Comparative analysis

Comparative analysis brings our knowledge and experience of our listening environment and its monitor path together to assess the overall tonal balance and dynamic range.

The use of audio referencing will really help in starting to conduct this type of analysis. But when you are very attuned to your workspace, there is less need for references to assist in this process. Though sometimes a cleansing of our audio palate is just what is needed to bring a fresh perspective. The action of referencing will facilitate comprehension of the environment and monitor path sound, enabling the engineer to ascertain what is a neutral balance or ideal/flat. Remember that a reference sounds great in all environments, but if you have engaged in this a lot, you will not require them as much to support your approach.

I often find it useful to visualise this overview of the reference ideal as a flat line of frequency from 20Hz to 20kHz. Our analysis then draws around this across the spectrum to give an appraisal of the actual frequency imbalance of the source mix in relationship. This principle can be observed in figure 5.3. Remember the horizontal line 'flat', our reference, is not technically linear in its gain structure across the frequency range, but it is an easy way to visualise how our mix under analysis differs from our reference in general. All our references are this horizontal line. If the mix under analysis is dull comparatively, it will sound dull against any of the audio references it

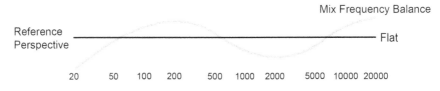

Figure 5.3 Visualised frequency difference between a 'flat' reference and mix source

is compared to. It is not about 'copying' another master, but about utilising the good tonal balance of other masters to help guide your own mastering in the correct direction to achieve a positive frequency balance within your master.

When playing all our references at equal loudness back to back, you will not notice one as being dull or bassy. You will notice the music is different and the tonal balance is not the same between them because each is its own unique piece of music. But they all exhibit a balance in general tone. This is what the horizontal line is – this average 'flat'. In bringing our mix under analysis into this average by appropriate processing, it will exhibit all the same positives in frequency balance on average but retain its own musical uniqueness as our audio references have.

In addition to this observation of the difference in our mix to the reference in tone, a visual appraisal of the dynamic of the mix is required to substantiate the tonal difference as momentary or static in nature. Observe if the tonal difference is moving in amplitude or fixed, static or momentary at any point on our tonal difference curve. This principle can be seen in figure 5.4.

This observation of the frequency balance of a static or momentary should not be confused with general dynamics of the whole frequency range, where everything is just too dynamic or too compressed.

Be aware that all this comparative listening is not the same as looking at a meter. Humans are amazing at analysis, there are no current analysis systems that give even a small insight into the above. If there were, logically, automatic mix and mastering algorithms would sound amazing every time. Luckily for the audio engineer's future, there is a long way to go before there is any notion of this coming to fruition, though the area of research around semantic audio is very interesting indeed. Semantic audio is the attempt to derive meaningful data from music. I am sure this will become more

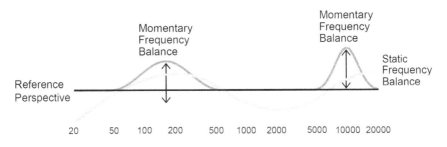

Figure 5.4 Momentary/static differences between a 'flat' reference and mix source

supportive of the engineer of the future to make using tools quicker and more musical. But it will be the user defining the approach, as we are the ones that can actually listen!

Human beings are fine-tuned listening machines developed over millennia to survive and communicate. Our ability to perceive detail within the general chaos of noise is unsurpassed because of our selective attention to the aspects in need of our focus. This does not mean metering and analysers in general are not helpful. They can be supportive of varied sections of the evaluative process but should not be the driver of any detailed audio analysis. Remember machines cannot enjoy music, they can analyse it, but this is not the same as listening to it, which is something humans can instinctively do.

As engineers it is critical to put the focus on the best tools we have (our ears), supported by the environment, critical monitoring path and references (again an aural source) to assess the overall tonal colour and dynamic of our source mix and achieve comparative analysis.

Reference selection and use

I previously outlined the premise of listening to superlative audio to help comprehend what a quality sonic outcome is. In practice, during mastering there is a different context to how a reference would be used to support mixing. In mixing, often the focus is the genre, such as setting the sound stage, the mix balance and gaining interesting stylistic interplay in the mix from a reference can be useful. In mastering, audio referencing has two clear uses.

One is helping the engineer comprehend the listening environment by observing the bass, middle and treble of the reference. In doing so, familiarise observations to the sound of the room and monitoring with the known sonics from the audio references. In tuning our ear to the correct impression of the tone of the room and monitoring, there is not a correct room or monitor, but there is a critical need to comprehend the sound of the room to be used.

The second use is making that same tonal observation and impression of the sound balance but comparing it to our session mix or final master to give an objective comparative, especially in tone. When engaged in this, genre is not the consideration, but the quality of the sonics of the audio reference. This concept can be quite difficult to become accustomed to, but the ability to detach from genre is an important part for your development towards being critical in analysis and able to observe overview. This is an important transition from wearing a mixing hat to that of the mastering engineer.

In comparison, informative observations can be made, for example, in listening back and forth between the mix and an audio reference having already balanced for equal loudness. Listen to the stereo elements where the mix aspects are panned or what is referred to as the stereo impression. Switch back and forth (A/B'ing), only trying to observe the change in width if there is one. If the mix sounds narrower, knowing the reference has a positive stereo impression means our mix under analysis must lack stereo width. If it is wider, then it has too much stereo impression and so on. In conducting this constructive comparison, the genre is not in question, just the observation of the stereo impression comparatively.

This non-musical based comparison can equally be made for the tonal balance or timbre of the track, asking if the mix sounds brighter or duller against the reference. The same is true with the depth of field or also referred to as the spatial impression. Lastly for sound balance, ask if the levels of the main instruments are too forward, and if the vocal is too far back in relation to the mix. Refine by asking if this is a tonal and/or amplitude shift. Don't try to make a musical observation, but a simple comparative observation around the difference between our reference and our mix.

You could argue that all songs are different, which on the whole they are, but if you conduct this simple analysis with fifty or a hundred sonically superlative references, you would find there is very little difference between them in this observation in overview, as they all have excellent control of the stereo and spatial impression, the sound balance and tone. This will translate well on all systems; they sound mastered, not exactly the same, but they do not sound wholly different from each other either, though the music obviously does. This is why it can be challenging to do this type of analysis until you become familiar with the premise.

The ability to ignore the genre/musical content and observe the overview is also helpful during comparative analysis when assessing tone. This will help to not only comprehend the reference's good tonal balance, but start to observe what is incorrect with our mix or master outcome. The more quality outcomes observed that have excellent translation on all systems, the more these commonalities of tonal balance are appreciated across genres. It is this average being listened for, using this library of listening to help our masters join this positive sense of translation. After all, you would hope your music does not sound out of place when played relative to any other quality song in the same context. In the same way, the consumer would not expect to have to turn the volume up and down during a track on an album as much as not having to reach for the tone control between different songs on a playlist.

You could argue at this point that mastering in some genres like 'classical music' is almost a completely different matter. Classical music should and can be captured perfectly in place, translating the listening experience of that actual performance in time and space to the recording for playback. There should not be a need to process unless this capture is imperfect. But either way, all the same principles in 'normal' mastering apply. If you listen effectively to the recording, you would not change anything – this is mastering. Making classical music an important part of your listening context whether you are mastering drum and bass, or metal or folk is crucial. This is all part of our library of listening. Being able to observe an overview of the sonic comes from stepping back from our prejudices of genre to hear the qualities of the music in both tone and dynamic. This is not easy to achieve; the more you listen to different genres, the easier it will be to do this with the genres of music you emotively enjoy. It is about detaching your perspective from the music to the sonics. Mastering is about observing and changing those sonics and in any music genre. This is in part why some people refer to mastering as a 'dark art' because they are not listening for the correct sonic changes made by the mastering process making it seem 'magic'. Obviously it is not magic, just correct critical listening for the task at hand. The more music and genres you listen to, the more you will hear this context.

Corrective analysis or an observation in overview?

Corrective analysis is the observation of a mix's tonal errors and the inspection of these to note whether they are momentary (dynamic) or static (fixed). Tonally, any anomalies are what is being looked for, underweighted elements or resonant aspects. Singular static hums and buzzes are unusual in modern recording, but are often artefacts encountered in restoration. More commonly it would be bass notes, kick or guitar parts with dominant frequency in the chords or melody structure. These are caused by either issues with the mixing environment not enabling the engineer to hear, or an engineer correcting an aspect because of their environment rather than an issue with the original source. Equally they can just be down to a lack of correct control of dynamics of tone in the mix.

These corrective observations are opposite to an issue with a mix in overview, where the aspect is not specific, but observed broadly across the music, for example, if a mix was observed as being dull or had too much bass. This would have been initially observed during comparative analysis against audio reference material.

Corrective aspects are normally observed in the individual instrumentation, for example, an excessive sibilance on a vocal or an overly weighted kick drum relative to the bass. These are often outcomes not immediately obvious to the mix engineer. They can be caused by the environment, but often the monitors used during the mix just lack transient response. These transient errors can be sharp and percussive in nature, for example, during the mix in trying to shape the snare to give it more attack, bringing it to the foreground in the mix. They cannot hear the sharpest part of the transient because the monitoring is not translating the attack response. On analysing, in a more critical monitoring environment, the snare pushes much further forward dominating the foreground. Also with bass, the monitors or amps can often compress the bass end, hence the dynamic issue in the lows are not obvious. None of the above is obviously the intent of the mix engineer, but the corrective analysis undertaken makes it clear these need to be addressed in the outcome.

At this point, it could be suggested to contact the mix engineer to correct, but what you cannot hear you cannot fix. Sibilance also often comes into this as an outcome, as the mix engineer cannot 'hear' how aggressive the hype of the treble is. Monitors are often to blame; if the mix engineer cannot hear the issue on their system, it makes it impossible to correct an issue they are not observing. Asking them to correct it can make the scenario worse. It is important to be considered in requesting amendments to a mix. Critical suggestions around the cohesiveness of a mix or its intent are more helpful than specifics.

The approach and contextualisation of this type of analysis is explored in-depth in Chapter 9 'Contextual processing analysis'.

Summary

In conducting in-depth analysis, note the dynamic and tonal outcomes through comparative analysis with audio reference material and any potential pathing of processing from constructive analysis, LRMS. Issues in the mix balance should be explored with corrective analysis observing whether an outcome is momentary (dynamic) or static (fixed). In doing so an encompassing impression to support development of processing will be created.

Notes about approach to stem mastering

There is a clear demarcation when engaging in analysis of a stereo mix as the mixed outcome is fixed. The mix is the mix. When loading stems, even though they are the composite parts that sum together to make the same mix in principle, the fact that the mix balance can now directly be changed psychologically affects the way analysis is considered as a deeper dive into the construction of the mix can be conducted rather than observing just the stereo mix in overview. I feel this stem analysis can be a slippery slope, where it is easy to become lost in fixing someone else's production based on our musical aesthetic, and not focus on the bigger picture of the sonics. For me, the final stereo mix should be signed off as an outcome from the artist, producer and label.

Practically, even if the client supplies stems, I always work with the stereo mix first. Only looking towards the stems if a problem cannot be fixed in the stereo construct, or else it has been noted after the first master reference has been supplied. Even then, with minor fixes such as vocal levels, I would rather create with a sum/difference of the source stereo mix against the instrumental supplied to extract the vocal and put this outcome with the instrumental back in the transfer path. A tweak to the vocal level can be applied, but it has been done so purely with the aspects from the final mix, rather than addressed on the stems which may not sum/diff against the 'final mix' source supplied and signed off by the client.

There is also the issue of 'glue' as soon as the stereo mix is processed as stems. If the main vocal was observed in the analysis of the mix to be sibilant, a split band compressor would be applied. Its trigger level would have been derived from the best response to the aspect that could be heard, but there would undoubtedly be other triggers. Points from the hats, ride or top of the guitar or synth may have dynamically strayed above the threshold set for the sibilance and will also be affected by this process. I would not view this as a negative, but part of the 'glue' of mastering being applied to the mix, because anything in that frequency range that went above the threshold of the dynamic would be affected. It breached the threshold because it was too loud! The temptation with stems would be to place this process on the discrete track/stem, thus missing out on the 'glue' that could be achieved by looking to master the track and not mix it!

I feel there is a breadth to the perspective that is lost when starting from stems and not just the final mix. Looking at the stereo mix is a more effective way to develop your ability to think in a mastering perspective and not blur the line between the mix process and mastering.

Mastering tools

The first consideration in selecting any tool for mastering is that 100% of the original signal will pass through the process unless utilising a parallel path. This means 100% of the tool's 'sound' or 'colour' will be imparted onto the source audio. When mixing and inserting a tool into a channel path, it is only discreetly applying its 'sound' to that individual channel in the mix, one of maybe 30 or 40 differing channel paths being summed to the master bus.

However, when inserting a tool in the mastering transfer path, it is the same as inserting that process on every single channel of the mix. When you start to think of it like that, it becomes clear that even a small change in 'colour' imparted by the tool can have a big effect on the sound overall. It also helps to explain why mastering tools are some of the most expensive in audio production because those signal paths are absolutely critical to the overall outcome.

It is not a bad thing that equipment changes the sound, it is just important to understand our equipment in-depth and have clear knowledge around what they impart from thorough testing (see section on 'tool testing' in this chapter).

The 'sound' of the tool type outcome is an addition alongside the outcome from the process itself. The tool's processing purpose would be referred to as the primary outcome, i.e., a dynamic process of a compressor or frequency filter of an EQ. The effect of the dynamic on the tone because the waveform has been manipulated is referred to as the secondary outcome. For example, downwards compression with a fast attack would dull the tone. This must be accounted for in the overall transfer path design of a given master.

Knowing how a tool changes the sound means it can be used to our advantage in devising a transfer path. You may not have considered it, but

DOI: 10.4324/9781003329251-7

even digital tools often change the overall sound of the signal before any adjustment has been made. These tools in general are trying to mimic analogue, and in my experience can sometimes have too much of this 'analogue' sound compared to the original processor. This makes it even more crucial to test and evaluate before use. It does not mean these tools are bad in any way, they are just different and require consideration in their own right. They should not just be used as a 'copy' of the analogue original. The same logic applies the other way around for a tool used extensively in digital where its primary, secondary and tool type outcomes are understood. In purchasing the analogue version, it should not be assumed it will work in the same way musically or sound exactly the same. It may not even achieve the desired outcome, just a different outcome.

Every tool whether digital or analogue has a tool type.

Tool type

In testing tools in both domains, it becomes clear how much sound is imparted on the signal path other than the original intended process – the primary outcome, equally how transparent or how many artefacts a tool has.

Often in mastering, a mixing tool is far too coloured in its sound to apply to the whole signal. The required outcomes are very different. In mixing we often want to 'change' a sound, imparting a uniqueness to that aspect of the mix to bring it to life. In contrast, mastering takes all the hard work of the artist, producer and mix engineers and sculpts the outcome to make it fit for purpose overall. It is crucial to be sympathetic to the aesthetic of the mix and not over 'colour' with our tools detracting from the original production.

When using a tool that delivers a lot of spectral change to the signal path, it is clearly going to impart that on the whole mix because 100% of the signal has passed through the unit. This is why there are very few real 'crossover' units between the mixing and mastering environments. The needs are very different because mastering tools on the whole need to retain the bandwidth of the original signal and impart minimal noise. For mixing this can often make it too unreactive or passive in use. This obviously makes mastering units more costly to produce by requiring a cleaner, more considered path in the analogue domain.

Careful research is required when considering a given tool. First, the domain should be observed. If it is analogue, inherently it means it will do something to the sound that is often the whole point of utilising it. This is called the 'tool type' directive of that processor. As an example, a mastering engineer would buy a Manley Vari Mu because of its unique musical response in downward compression (the primary outcome), the overall aesthetic is dulled (the secondary outcome) and the dynamic range is more controlled (reduced in contrast). This dulling/secondary outcome affects the tone, though there is also the 'tool type' to consider because it is driven by valves.

Often also referred to as 'tubes', they are one and the same. Their names come from the vacuum tube (valve) being equivalent in function to today's modern transistor. This is also a switch/regulator, or can be thought of as a valve that controls the flow of input to output as a tap/faucet would. Again, they are two words to describe what we all see as the same thing. The simplest form of vacuum tube is a 'Fleming valve', this is equivalent in function to a modern diode, hence the term 'vacuum diode' is also sometimes used. You will see the terms triode, tetrode, and pentode which evolved the design of the simple vacuum diode. This is a complex area and all valves are not the same! But in our context of looking at the whole family of this circuit type and its attributes, the terms tube and valve are commonplace and both are correct. I have grown up using valves as a term, and will continue to use 'valve' throughout the book.

This valve circuit type in the Manley Vari MU imparts mild saturation in general use and by consequence generates harmonics. Making it often brighter at the output than the original input signal even though it has been compressed or levelled, remembering that the secondary outcome of downwards compression is to dull the tone overall. The primary outcome is that it compresses, which a Manley Vari MU does in a distinctive way, as it uses the valve to apply gain reduction which also adds to the unit's appeal. But equally, some engineers will use the unit as part of the path just as a soft drive with no gain reduction, switching the 'tool type' outcome to the 'primary' outcome and using the unit as a saturator. The more a tool is investigated through testing, the more effectively and creatively it can be used in a mastering context.

If the domain is digital, all the above could be true when utilising an analogue modelled compressor, the Universal Audio Digital (UAD) Manley endorsed Vari MU being a good example. It will give a similar outcome in gain reduction, but is not the same in tonal colour because it is more active in its saturation. The different units should not be confused between analogue and digital, or equally get into an analogue versus digital debate. The overriding important aspect is to understand the primary, secondary and tool type outcomes of each, knowing when either would be useful to use in our creative process, not which one is 'better' than the other. Analogue is not digital and digital is not analogue, and these two domains should not be conflated together but used where required to achieve the best outcome for the given scenario.

The circuit type of any analogue audio device will do something to the sound if it is processing, whether very slight or clearly noticeable. Some digital tools will do nothing until changed (clean tool type) but some will impart noticeable distortion and/or tonal shift. The more different processors are engaged with, the more skilled you will become at using these tool types to your advantage. But to do this effectively, it is critical to engage with tool testing.

Tool testing

As we remember, louder really does sound better. If you were a plugin designer, would you not want to use that knowledge to your advantage to 'sell' your products? You can make your 'analogue modelled' process sound better to the user as soon as it was inserted in their DAW, and make the sound perceptively 'richer' or 'warmer'. Simply making it a little louder achieves both these outcomes when bypassing on/off. Not surprisingly, there are many plugins that do this very thing. Once inserted the default setting is just louder than when bypassed. It may also be changing the colour of the sound, but the effect of the loudness difference is what makes it stand out to the uninitiated user.

Some do this apparently with a good rationale as the unit being modelled imparts a tonal colour when passing signal. There are examples of this 'louder' outcome in the analogue domain too because louder has sounded better for a long time in the world of audio engineering. This does not mean it is wrong, just that there is an amplitude difference at input to output.

None of the above actually matters if adherence to correct engineering practice is observed and a re-balance for equal loudness is actioned. The choice is ours as engineers to correctly test equipment to discover what artefacts there are within a given tool. There are no excuses for not knowing how a specific tool affects gain structure as well as any secondary or tool type outcome. Once this is fully comprehended, these outcomes can be used to our advantage fully without any bias from loudness.

If a tool adds harmonics via saturation, like a valve process, whether this be a real valve in the analogue domain or a modelled one in digital, it is making the signal more present and will make the overall outcome brighter. If the source mix is already harsh, this tool type would be a poor choice no matter what the cost or quality of the unit. Conversely, if the source is dull, the added excitement will contribute to the overall aesthetic making less processing required in our considerations.

Often to achieve a minimal processing path, the tool type outcome and the secondary outcomes of the actual primary processing are required to reduce the need for multiple processes. As the saying goes, less is more!

In digital, it is very easy to assess a plugin's transparency or to ascertain its artefact by simply using sum difference.

Sum difference

A sum difference test is a simple comparative exercise between two of the same sample aligned sources to measure the difference (diff) between them, often because one of the sources is modified by a process or change in the path. It is clearly an effective method of evaluating difference. This is the same principle as used to create the 'Side' observation from a stereo mix – left plus right, with the right polarity reversed and summing to a mono bus.

When utilising the same principle with a stereo file, the sum is against a copy of the original file, which will create a null but only if the paths are identical. But now insert a plugin on one of these stereo channel paths, assuming the delay compensation in the DAW is working (you may have to start/stop to reset it). This delay is the DAW calculating the time difference caused by the plugin using the digital signal processor (DSP). Once 'in time', if there is any sound present, then the plugin is making some alteration to the sound in its default loading. With some plugins, like a clean digital EQ (with no gain or attenuation applied) or a digital compressor (with ratios of 1:1), nothing in principle should be heard. If it is truly transparent, the sum/diff will contain no signal. It is easy to assume digital is transparent as there is no obvious noise. But if an analogue modelled tool was inserted, it is likely there would be a difference, this would be the change by default the tool is applying.

In hearing this difference, first it needs to be established if there is also a loudness difference as well by adjusting one of the channel faders up and down in small steps (0.1dB) to find the point where the difference sounds the smallest – closest to null. The dB difference evidenced on the fader is the volume difference. This may be 0dB but equally it is not unusual for this to be at least a 1dB.

When A/B'ing between the original and the process, the true difference can be heard without bias from the loudness change. Listening to the difference can help our ears discern the sonic changes taking place. When observing the difference, it is only known that they are indeed different, while it is not known if the sonic heard is positive or negative in amplitude, just that it is not the same at those points in the frequency range. Often in switching and referencing back and forth once out of sum/difference, it is possible to ascertain if the difference heard was a treble boost or cut, mid presence peak or bass boost. When 'hearing' the change in detail in sum/difference, often it is easy to pick it out in an A/B, when initially it was difficult to discern what was happening across the frequency spectrum.

Having had the opportunity to 'tune in' to a sonic difference, it is clearer to hear in complex program material. This is the way with any sonic analysis, being able to use an analysis method to gain focus on the source. Once heard, it becomes easy to pick out in the big picture.

Two channels are better than stereo

LRMS analysis facilitates engineers to break down the composite parts of a stereo mix to better understand where differing parts are placed in the image, frequency balance and dynamic, part of what is called constructive analysis.

When applying a process, our tool selection should also allow targeting of these LRMS paths. This is the basic rationale for why all professional mastering tools have two discrete channels, often referred to as 1/2 or A/B (not to be confused with A/B comparison or A/B scenes for recall). Two discrete channels means processing can be applied to the left/right or Mid/Side aspects of a given source. Targeting it in the most effective position in the stereo image. The same then applies for the sidechain (SC) of a dynamics tool. Having discrete channels enable the M to trigger the S or the other way around. L/R can be dual mono split on rare occasions where the mix is hard split in its left/right relationship. Normally in stereo using a linked SC would be appropriate to avoid the left/right image becoming unstable with differing gain reduction on each side. Targeting the stereo as a whole is fine if processing is required in that aesthetic.

To truly realise the power of our processing tools, it is a given when first approaching analysis that a degree of precision is needed to take account of all aspects of the stereo image using LRMS to break it down. It is not random that dual mono tools have been the mainstay of mastering engineers since the development of stereo. The earliest stereo compressors used for cutting to lacquers were often denoted as in Lat/Vert mode as opposed to stereo. This refers to the lateral or vertical motion of the stereo cutting head. A mono cutting head is only lateral in motion. Lat/Vert is now referred to as Mid Side or mono and difference. This enables the cutting engineer to control the mono weight of the lateral motion of the cutting head evening out the signal feed to the pitch control unit which achieves the best groove spacing. The vertical (Side) dynamic range can be controlled to avoid lift off or break through the lacquer during a cut. I cover these and other principles of control required for cutting in Chapter 16 'Restrictions of delivery formats'.

Discrete channels enable the mastering engineer to make a direct link between analysis of the music and the path of the process. How that processing is enabled is enhanced by step controls.

Step controls

Step controls are a feature you see on all professional mastering hardware and software applications on the whole. This is due to the outcomes step controls impose on our methodology in approach. These are: recall, A/B comparison, audio quality, benefit of fixed frequency and numerical technique in approach. All these aspects are discussed in the following sections.

Easy recall of hardware. A fixed position makes it straightforward to log (manually make a record of) and accurately recall changes required at a later date. A variable potentiometer is not as stable in connectivity or recallability of position as a fixed step switch position. A step control is also better for conductivity making for superior audio quality in analogue. This also enables discrete circuits for each path with a stepped control creating the conditions for more accurate channel matching by removing variable elements. When working in stereo, this minimises negative effects on the stability of the image because channel 1 and 2 matching is crucial. This is straightforward in the digital domain, but requires consideration in any analogue design. Overall, the fixed element improves the stability of the unit and its precision.

Even a very high-end mastering tool in analogue is not perfect, especially in recall. It is very close in sound, but for instance a valve changes over time, and a recall a few months down the line will not be identical. Even the room temperature can affect it, which is why it is best to moderate the control room temperature. It is not just the people in the studio that require an air management system but the gear also!

There are ways to mitigate left right image imbalance with analogue gear; the simplest is using it in Mid/Side. Any discrepancy in amplitude or tone mismatch between channel 1 and 2 can be compensated on the output without disrupting the stereo image. It is not unusual for valve gear to have an acceptable tolerance in manufacture of 1dB between channels. This is quite a lot of volume!

When using analogue, investing time in knowing your gear is vital to make sure a tool's positive and negative practicalities are fully understood. The quirks of individual units are generally not an issue in the digital domain but these processors still require thorough testing to comprehend their sonics.

The benefit of fixed frequency

Having fixed values means an equaliser has set interrelationships between frequencies across the spectrum, meaning it makes sense that they would be musically sympathetic. Allowing the frequencies to pocket with each other and not overlap incorrectly makes the practical use of the tool very quick because choices are limited yet equally musical during the mastering process itself. This pocketing is true of digital tools too. TC electronic system 6000 EQs or Weiss EQ1 use this same 'pocketing' to good effect. If you look at a professional mastering tool setup and its frequency sets, make a listing of these so you can use the same frequency points on your variable EQ to benefit from the interrelation of frequencies. It will make it more musical in a mastering context. Remember, mastering is about broad brush strokes – the fact that it is not possible to tweak from 800Hz fixed position to 820Hz is not an issue. It could be restrictive in mixing though maybe that is one of the rationales for mixing EQ to be variable. Selecting the right tools for the task at hand is helpful.

This fixed element in switching from one position to another means it is easy to create a clear A/B comparative.

A/B comparative

Switching from 1.2k to 1.8k with a stepped EQ, the observation being made is strictly comparative in listening to the difference between those two frequencies. The change is a clear switch and not a sweep over time as with a variable potentiometer. This is the most important concept to take onboard from practical engagement in evaluation with set values. Variable controls are useful for 'finding' a musical frequency to target in mixing, but often in mastering it is ideal to make broad brush strokes across the frequency spectrum. The target is often less focused in initial analysis, and in evaluation the focus is made with the correct position in an A or B decision. For example, in initial analysis observing the mids requires more focus in frequency around 1k, actually in evaluation it may be 1.2k, 1.8k or 800Hz on our fixed position control. Having the steps means changes and decisions are focused, unlike with the EQ making variable sweeps to 'find' a frequency range. When using a fixed position, a clear A/B comparative is being made. One of the rationales for why many digital tools have an A or B setup option is that it enables this same clear comparative between outcomes or whole plugin setups. Most digital plugins have an A or B comparison mode as part of the generic plugin interface if the DAW allows this. Using these facilities will enable a more effective evaluation of the application of processing and result in a better outcome for the audio processing overall.

In A/B'ing the fixed values we are not sweeping variables between potential points. Our focus in listening is between the exact selective positions. This makes us consciously read the step positions value, meaning we are using numbers.

Using numbers

Stepped application principles should become an integral part of our methodology even with variable control plugins in our approach to manipulating audio in mastering. In using a compressor and listening to the attack, it is helpful to be entering values to observe change. For example, after analysis of a mix, you may decide to use a slow attack at 50 ms as a starting point, which would be appropriate, but in evaluation it needs more bite. Typing the numeric in at 40 ms while the audio is playing means it is a clear change, and the step, the difference is observed. Switch back again by typing 50 ms to observe the change again, the same way a stepped control between 40 and 50 ms would work. Sweeping between 40 and 50 ms is a transition between those points and not the difference that is heard. This makes it much harder to evaluate the difference quickly and effectively in overview. When listening to complex program material, it is the aesthetic change overall that requires focus, this is very different from listening to a vocal in the mix while compressing and dialling back the attack to hear where it opens up. In mastering compression, there is a complex set of transients to observe how they are changing. Clear A/B between control positions makes this observation more critical. The same principle would apply with an EQ gain or frequency, listening to the difference and not the variable sweep.

It is worth considering this approach in a mixing scenario where it is clear why the application is being made, for example, with a bus or vocal compression where a potential change has been assessed and a rational has been evaluated towards approach. But equally it is fine to be more experimental with EQ on an instrument to find an ideal range to cut or boost, especially if you are unsure of what you are looking for aesthetically. The same applies with dynamics if you are being experimental. By the time the mix gets to mastering, this experimentation is all done, making the comparative approach vital for the previously discussed step rationale in A/B comparative. There is also an association developed with the values when using this methodology. In our compression example, the values used in focusing the attack with that specific tool on the mix helps us to relate to subsequent occasions in the same scenario using any compressor, as this numeric information is transferable. Logical interrelationships between numbers also enables the link between the sonic and numeric, when building a library of these values to reference, especially when first setting up a tool for use in a chain. Setting a tool with approximately the right values before enabling in the path means it should sound something like intended straight away when inserted. Minimise your contact time with the audio to make better decisions focused around your initial analysis. This makes it easy to assess and tweak rather than hearing the tool change the sound from the default with the music playing, until we have made it somewhere near what was intended. That is all wasted critical listening time.

This is a key aspect in using all tools when mastering. Noting all values as you go and logging these in your mind builds up a clear picture of the tools' relationship in dynamic and frequency control with the processing used across the whole transfer path. This is especially true in decibel relationships, where noting threshold and gain reduction in decibels makes the relationship obvious. The numbers are being actively read and we are not just looking at the graphics.

Not only does this improve our ability to make 'tools' work together and relate to the audio in question, but it also builds up a picture of how each tool responds in its control range. With tools, a value can be defined, but it does not mean it is the same as the same value on another tool. Making notes helps us to start to evaluate the character of each tool, its knee shape, slope, recovery shape, etc.

The constant adherence to noting values will vastly improve awareness of pitch to frequency in analysis. Listen and find the required frequency outcome and consciously observe the value. Make the link between listening and the actual frequency range. The same applies with dynamics, and in some regards this is the harder skill to achieve a high level of proficiency in. Noting ratio value difference especially will help you to start to 'hear' the compressor more effectively and become more considered in application.

Tool types, primary and secondary outcomes

The primary outcome is the most obvious change to the audio. For example, a compressor primary outcome would be to reduce the dynamic range of the audio. An equaliser would be to change the frequency balance and so on. Though as discussed previously in the chapter, a valve compressor could be used purely as a drive to add harmonic distortion to the path. This would normally be the tool type outcome, but in this application, it becomes the primary outcome as no compression is applied. If applying compression, this becomes the primary outcome; the harmonic distortion in the signal path becomes the tool type outcome. But there is also the effect of the compression on the audio, the compressor's secondary outcome. When applying downward compression, the dynamic range is reduced as the impact of the transient peaks are also, even more so during limiting, but both would give an impression of reduction in the treble, thus dulling the audio. In achieving upward compression, the dynamic range is also reduced, but this time the lower parts of the waveform are increased, making the sound more sustained and increasing the impression of bass. Upward expansion lifts the transients, and doing so increases the dynamic range and our perception of treble. The secondary outcomes must be considered in the path along with the tool type and the primary function. The compound effect must be taken into account when designing a given transfer path. As an example, a linear phase equalisation will not impart any 'tool type' or 'secondary outcome' being a clean, cold or sterile process, it just makes an amplitude shift in the given frequency range. This EQ was then made dynamic in compression response and was changing a broad range of frequency, say expanding the top end. As this is now lifting the transient, it is also creating more perceived treble from the secondary outcome, hence the amount of EQ could probably be reduced in gain because the action of its movement in contrast on the audio is more obvious.

There are many potential combinations of outcomes with any tool process and tool type. Consideration is needed for all in our transfer path design to make sure everything works in sympathy. Make sure that tools and processes are not fighting with each other's outcomes.

Summary

Understanding the consequence of 100% of the signal going through the selected mastering tool in our transfer path is critical. This tool has a tool type, whether clean or imparting colour in tone and it must be understood. The tool should have two discrete channels and the controls are stepped, or are used in a stepped configuration to action changes. We should have thoroughly tested tool type outcome, primary functions with controls and any potential secondary outcomes. Acting on this, our ability to master practically and also use theory to our advantage is significantly enhanced.

When I first started engineering in the days when digital was just coming to the marketplace, nearly all our tools were analogue. When acquiring an audio device, the first thing to do would be to test it. Because it is analogue you would know it is going to do something to the sound. Even with early digital tools the same applied as signal to noise from ADDA was often an issue that needed to be understood to achieve the cleanest path as much as would be done when using tape. In our modern paradigm and high-resolution digital world, I feel this principle has been lost or not taught to up-and-coming engineers. Even more so now than back when I was first starting out, I think it is crucial to test. The differences are now vast in our new digital hybrid analogue copied world. This copy is not bad in any dimension, I fully embrace it and I am glad of this technological progress. But testing to fully comprehend our tools is as crucial now as it was then.

Equalisation and tonal balance

Throughout my career, I've become more and more aware of how over-used equalisation (EQ) can be. If I look back at my work in previous decades, EQ would feature in the sonic outcomes more than it would in my current paradigm. In part this is because the quality of mixes I receive has improved. Maybe this is because monitoring quality improvements have enabled engineers to enhance their outcomes, as has access to better quality processing control in the DAW. Even the microphone and preamp quality to cost ratio has dropped significantly since my first adventures into recording. But the majority I feel is my appreciation for how powerful dynamics can be in shaping the tonal colour and unifying a mix. I also recognised that some aspects I thought I needed changing did not actually need to be addressed. Leaving a mix to stay as produced and no longer applying corrective changes derived from how I thought it should have been balanced.

In the same way, as with my use of EQ, I can see a clear shift away from multiband processing to full band. Doing so achieves a deeper unification of the original mix balance. Out of this, I also have gained an appreciation for the power of EQ, and what it can bring to the outcome of a given master if used subtly, rather than what I would now see as an exuberance to control the mix amplitude with EQ over utilising full band dynamic control.

Before engaging with EQ, there are simple principles to frequency balance that must be observed at all times. When changing one aspect of the frequency balance, another will move in our perception in response. This principle affects all our actions across the spectrum. In increasing the bass there is less high end, and reducing the lower mid, the upper mids will become more excited. Every action has an effect, and this yin yang of EQ

DOI: 10.4324/9781003329251-8

needs to be at the forefront of our thoughts as much as equal loudness when manipulating amplitude difference.

To EQ or not EQ?

Engineers in their formative experiences often initially embrace EQ as it is self-gratifying; in applying even small amounts, the difference can be clearly heard. Dynamics on the other hand are much more challenging to hear until something is obviously wrong or clearly being changed. This potential habit of overuse needs to be undone to get the most out of EQ in the mastering process.

The more I explored the use of compression as a young audio engineer, the more I became aware of the link between dynamic movement and our perception of tone. As discussed in previous chapters, an increase and decrease in amplitude changes our perception of tonal balance. A dynamic sound getting louder sounds fuller/brighter, and a quieter change leads to our perception of tone as softer/duller/weaker. Often this change is in the transient details and is not obvious when listening to a passage of music that is complex program material. The brighter/louder elements of the transients fool our impression of the amount of treble in the balance overall. A good example of this is a sibilant vocal in the mix; the strong burst of treble distracts our observation of the tonal balance, leaving us with the impression there is treble overall, or a constant hi hat in an electronic dance music (EDM) mix can make it feel like there is a positive treble balance. Though if it was possible to mute the hat in the mix, everything would now sound dull because our hearing 'fills in' the tonal gaps when the strong hi hat treble or the sibilance is present.

This is in contrast to diffused music, such as many classical recordings capturing time in space, the environmental element is a key aspect to the delivery of the sonic overall. Making the sound naturally compressed in dynamic as the room diffusion fills in the gaps between the transients and the transient are less direct as generally further away from the capture device. The music's focus on dynamic crescendos delivers the musical emotive impacts. In this case you are unlikely to come across a scenario where the music suffers from these mixed transient tonal issues. It arises in genres where differing source recording paths or created sounds are mixed together, with focus on the discrete attack of elements rather than a singular diffused sound overall.

This bounce in dynamics is less obvious in the bass where the sounds tend to be more sustained, making the focus around the RMS of the signal. This can be perceived as a static weight whether positive or negative in balance. But the bounce is still there, causing masking in the sound balance. When a larger amplitude change is present, it will be noted clearly though sometimes more as a resonance in frequency. But it is just the dynamic difference in tone because of the loudness. When controlling these amplitude-based events, the true tonality of the song will appear; it is unmasked by the correct control of the dynamic balance. All this focus around dynamic control does not mean these processes should be used first by default, but it is important to comprehend that equalisation is not necessarily the first port of call when it comes to a response to an observation in frequency balance. The clearer the detail in our observations, the more confident it is possible to be about the actual change that is taking place. Hence it is crucial to observe whether the tonal change is static or momentary.

Static or momentary

To differentiate between tone that is static or constant and one that is variable is critical in our observation to link analysis to the correct process to manipulate. For example, a kick could have too much low end, but in overview, the mix is not over weighted. In this case, static EQ is not a correct response because the incorrect frequency observation is with the kick balance on/off, meaning the tool used should be able to achieve an on/off response. Once corrected, it is likely the whole tone would lack bass as the kick is fixed. The same principle is discussed above with the hat or sibilance changing the perception of the frequency balance. Remember that with EQ every action has an effect – changing one aspect in tone will alter our perception of an opposing aspect.

As technology has developed over the last few decades, it has brought the concept and application of dynamic filter bands, what is commonly referred to as either multiband dynamics or dynamic EQs, of which there are differences in their design and control. When something is fundamentally wrong in the frequency balance across time, a focus dynamic band of EQ is truly wonderful in addressing the outcome.

There is a straightforward principle to follow when it comes to selecting EQ when mastering. If the aspect to change only happens at different timing points in the music, it should be targeted dynamically because it is momentary. If the aspects are constant/static, a fixed gain EQ change would be most effective. An example is a snare that sounds dull relative to the rest of the sound balance of the mix but only hits on two beats in the bar. Upward expansion could be used on the snare to brighten those timing points. But if the snare, its reverb tail and other aspects in that area of the frequency range are dull, i.e., vocal, mid of the hats and so on, an EQ would be the most effective tool to address this broad change in tonal inaccuracy. All the frequency range requires more energy, not just the snare hits.

Dynamics are hard to hear and difficult to develop a high level of skill in application, meaning the shift from EQ to focus on dynamic control can be a long time in coming. Also, the argument that a multiband compressor being split frequency is EQ can be a negative influence. This is obviously correct, but the important part of the control is the dynamic aspect. I would rarely use multiband to shift the amplitude of the band gain range I'm processing, rather I would use the split frequency range to control the amplitude in that area because it is too dynamic or not expanded enough. The overall fixed gain or 'makeup' of the band should not change as would happen with full band compression where the makeup would be used to retrieve the lost amplitude. When correcting the given frequency range, there is nothing to 'make up' because the error is being brought back into line. Using a multiband dynamics processor to just change gain between filter band sections is not a positive approach to tone control in mastering. In mixing this is a different matter. This approach can often save a sound especially in the lower frequencies because of the aggressiveness of the filter shapes making the tool highly corrective, as discussed in Chapter 8 'Dynamic processors and tonal balance'.

Static observations should be responded to with a static EQ filter, and for a momentary observation, a dynamic filtered band is best. When trying to breathe life into a mix, a dynamic EQ is positive in an expansion mode. This is a combination of dynamics and standard equalisation curves making it twice as powerful, but equally two chances to ruin the mix in mastering. Focus is required in the analysis of the mix to correctly assign the appropriate tools to change the outcome for the better.

How does EQ work?

There are some basic principles in an EQs function that need to be appreciated to understand when and how these affect the sound when processing. This is a straightforward explanation to highlight the important sonic factors that need to be comprehended. If you are interested in the electronics or DSP coding involved, I suggest you look for publications in those areas to investigate that context. Our learning is focused on comprehending the sonic implication in the use of differing types of EQ tools and processes.

Starting by observing figure 7.1, the top path is a simple pass filter with a signal flowing through it. There are two things that must have changed at the end of the path (Output) relative to the start of the path (Input). One, the amplitude must be lower, and two, the signal is later in time. After all, time travel has not been invented and by putting a process in the way of a signal, the phase relationship between the input and output will have changed. This is why 'all pass' filters are used to make simple delays. In this case, the whole signal is delayed without any change in the audio spectrum from the filter, a frequency-dependent delay.

Add a parallel path in the lower graphic in figure 7.1, 'tapping off' the signal from input and routing it without any processing and summing the two paths together. The new parallel 'B' path has no delay relative to 'A', meaning at the point of sum there is phase difference between the two routings. This sonic outcome is called 'phase distortion' and the difference in time is described as the 'group delay'. You could ask, if this happens, why add this clean parallel path to the circuit?

The answer is in the filter shape achieved in the outcome from the sum. There is 100% of the original signal on the parallel path across the spectrum and 100% of the signal on the filter path above the frequency corner; this is the frequency numeric set on the filter. Below this frequency point there is a continuous reduction in percentage until it reaches 0% signal in the lower bass frequency. Overall in the sum there is now more treble than bass.

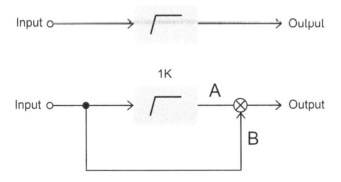

Figure 7.1 Simple EQ design principles

Analysing the new filter curve, a high shelf (HS) shape has been created, as can be seen in figure 7.2. This is a simple way to make differing filter shapes on a passive equaliser. In this example, the more or less of either path changes the relationship in how much treble would be heard. Flip the filter to a low pass and the summed outcome is a low shelf (LS) boost.

Phase distortion is an integral outcome in any EQs circuit design. Logically all EQ, whether in the analogue or digital domain, will have phase distortion. In digital, unlike analogue, it is possible to time travel as any latency (time difference) can be compensated for. I am sure that when the digital signal process concept was first practically realised, all EQ designers thought this was an amazing opportunity to design an EQ without any phase distortion. They did create this by making the group delay in time with the filtered signal across the frequency range. Remember that wavelength changes as the frequency does, meaning the delay is longer as the frequencies lower. Making this calculation is far from straightforward. The processing requires a high level of precision and DSP load to achieve the correct outcome. When the maths is perfect, the delay is linear and in time with the filtered signal in all aspects. This is called a 'linear phase' equaliser. There is no phase change introduced from the frequency gain change.

Now think of all the classic equalisers you have known or have heard countless engineers discuss. The majority of these would have been analogue, and must have phase distortion in the outcome. Phase distortion is not a bad thing, this smear in time is one of the reasons why different EQs sound the way they do. It is part of their characteristic. If designed effectively

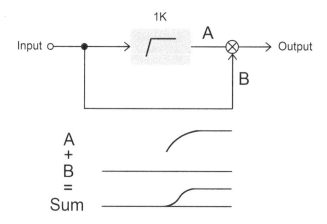

Figure 7.2 Simple passive EQ filter circuit and outcomes

this is a significant proportion of the reason why some EQ sounds active and others smoother, like a classic Neve channel strip EQ versus an SSL. The latter has less phase distortion in the circuit design and is thus less coloured. Both these EQs are loved by engineers around the globe in part because of the phase distortion. This is just as true for mastering EQ. Who would not want a Pultec in their rack, or a modern version, a Manley Massive Passive?

This observation means that 'good' EQ also sounds 'good' because of the phase distortion. Logically in making a linear phase EQ, this 'goodness' is removed. So, what does this sound like? I would assess its characteristics as cold, heartless, and clinical. All positive descriptives in moving the frequency range, without it sounding like any change other than the amplitude shift, have been made. Adjusting the frequency range clinically in gain, there is no tool type outcome at all. It is basically, a volume control for frequency range. This is what linear phase does brilliantly, but it has no heart, the same way your DAW master fader controls the output level. You would not expect it to affect the sound, but just change the amplitude and relative headroom at the output. All this means that it is critical to know the purpose of the application of EQ and what should that EQ shift sounds like in the aesthetic overall; like EQ, or just amplitude rectification.

This clinical outcome exhibited by linear phase is why it is referenced on some split band or multiband dynamics processors as an option. Because the filters used to split the frequency range for dynamic processing are 'in time', with no phase distortion, i.e., there is not a linear phase compressor. But if bands are created with linear phase filters to achieve zero colouration, these ranges can then be controlled dynamically. It is the filters that are linear phase, not the compression. Making this calculation less accurate reduces DSP load which would often be referred to as minimum phase. Some would say this makes the filter or crossover sound more analogue, but I would say this just blurs the lines in purpose of the application and hence have never found any use for it in a mastering context.

By using sum/difference, these EQ shapes and group delays can be clearly heard. The phase distortion will be in the sum/diff as well as the filter change. When using a linear phase EQ, the shape of the filter band selected should only be heard. Remember when listening, it is not possible to know if it is a boost or a cut unless that information is known from viewing the EQ gain setting. Without this information, it is only known there is a difference. This can be a powerful way to start to comprehend the different shapes of a bell or shelf and their Q with multiple EQs.

Series or parallel

There is another aspect to consider in the circuit design principles as most EQ has more than one module, generally three or four, bass/middle/treble and so on. Each of these introduces its own phase distortion when active. One way designers found to reduce this compound effect was to create the modules in parallel rather than series, as with standard mixing EQ. This difference can be seen in figure 7.3. It means each change in phase does not directly affect the next, making the EQ sounds less active from both the change in frequency balance as well as phase distortion. A parallel type circuit is generally positive for mastering, and is one of the differences between mixing and mastering EQ.

This parallel aspect also changes the relative gain structure because the amplitude moved is not compound, making summative shapes easier to create with a parallel circuit type. The cuts and boosts sound less like they are working against each other in creating an overall curve. Most mixing console EQs are series in design, in part because this is more cost effective, but equally as they are more active sounding, which is generally a positive outcome in mixing. Examples of hardware parallel mastering EQ would be Manley Massive Passive, SPL PassEQ, Maselec MEA-2 or as software the Sonoris Parallel or Kush Clariphonic MKII.

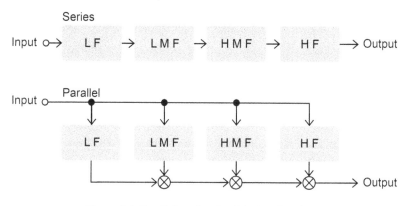

Figure 7.3 Parallel and series EQ signal pathing

The quality factor

The quality factor (Q) is the width of the filter curve around the centre frequency. This width is also expressed in bandwidth (BW) in octaves on some EQ. I prefer the use of Q in discussion but both are interchangeable. The higher the Q value, the tighter the width of the filter and the more focused/ resonant the shape. The opposite is true for the same expression of BW in octaves. For example: Q 0.7 = Octaves 1.92, Q 1 = Octaves 1.39, Q 1.5 = Octaves 0.95, Q 2 = Octaves 0.71, Q 4 = Octaves 0.36, Q 6 = Octaves 0.24. It does not matter what you use as long as you comprehend the width of the EQ filter being applied and its relationship to another.

As a general theme when applying EQ in mastering, broad areas of lift are the most effective in controlling the frequency spectrum. A bell Q set from 0.7 to 1.5 is quite standard, broad brush strokes. A tighter focus than this can often sound too intrusive. But when making corrective changes, a Q of 2 can often be helpful, which is a shape that would be more normally associated with mixing EQ. In mixing anything goes. Often it is the space between instruments that is trying to be sculpted, and tighter Qs are useful for this because they are for reducing gain without removing too much of the music. Sometimes, a tighter Q boost of 4 or 6, especially in the higher frequencies, can help to highlight musical accents. Although it is best to avoid high boost values with this as to not notice the EQ boost itself. Remember, the wavelength is shorter the higher the frequency, making our appreciation of the Q change across the spectrum. Normally highlighting an area would be across a slightly boarder range. For example, to emphasise the presence on a vocal, you could apply a lift as a bell at 8kHz, Q 2, +1dB.

Mastering's not so secret weapon – the shelf

We love shelves in mastering. One of the main rationales for this is a shelf's ability to only audibly change a portion of the signal being processed. This is one of the reasons why there are multiple shelves on a mastering EQ. In a lift or cut from the frequency corner, it eventually reaches a plateau, meaning the rest of the lift or cut is a near flat amplitude adjustment. Comprehension of the quality factor (Q) setting is needed to affect this correctly. This volume difference in the shelf can be observed in figure 7.4 where the volume 'lift' section has been cut from the full shelf shape above to evidence this simple lift in amplitude.

When adjusting the dB per Octave (dB/Oct) to the lowest setting and/or flattening the Q, a tilt can be achieved in balanced across the audio spectrum, like making a volume adjustment but across a frequency range. This is

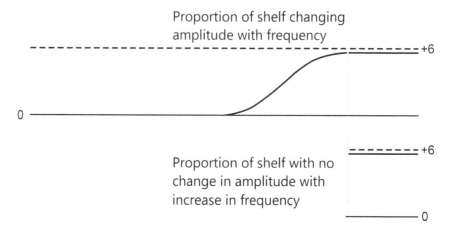

Proportion of shelf changing
amplitude with frequency

+6

0

+6

Proportion of shelf with no
change in amplitude with
increase in frequency

0

Figure 7.4 Shelving/gain change and not frequency shift

very much like the tilt of an old style hi fi tone control, but this would tend to keep rising or decreasing over frequencies.

Looking at the bell boost in figure 7.5, every part of the frequency range has a differing level of amplitude and phase shift. The bell shape has a focus point at the apex of the maximum gain, if pushed hard this could become more resonant and the change in frequency balance will be clearly heard. This is fine if that is the intention from our evaluation of analysis to highlight a frequency area in the mix. But a shelf only has this same effect on the area of transiting to its maximum gain. This is often a small part of the frequency range that changes overall with no focus point (resonance/peak) to the maximum gain, as can be seen in figure 7.5. This makes it sound a lot less like EQ is being applied. The plateau of the shelf is just an amplitude shift (see figure 7.4). The volume has been turned up in this section of frequency, not

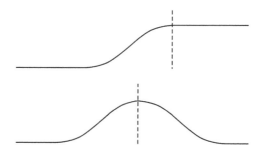

Figure 7.5 Difference between bell and shelving shapes

rebalancing the gain differently across the whole frequency distribution of the filter as would happen with a bell. The shelf sounds a lot less active/focused.

The focus exhibited by a bell filter is one of the main differences between mixing and mastering EQ shapes. A mixing EQ tends to have a sharper focus or resonance the higher the gain is pushed to drive focus in its use. The Q gets tighter the more amplitude that is applied. Mastering EQ has focus but maintains the Q when boost allowing for broad EQ lifts. The Q can then be used to change the focus if required. It should also be noted that mixing EQ often notches when cut, making the reduction in gain distinctly different from a boost. This is not the case with most mastering EQ, where the positive and negative change is often similar. This means it is clearly in your interest to comprehend the shapes of the EQ you are using. Read the manuals of the EQ, both those you own and do not own but are interested in, which will enlighten you as to how they change the frequency range relative to gain applied and learn important transferable knowledge in potential approaches to equalisation. The shape of the EQ filter is often as important as the processor tool type so you can in part emulate other EQ if you comprehend all these aspects.

In conducting this research, you would have also noticed nearly all mastering EQ has multiple shelf bands. This allows step EQs to reach a wider set of frequencies than just operating with the highest and lowest ranges. This extra detail in the mids allow for tilting from the mids out, and if another lift is required in the sub or air, another shelf is there to help shape the tone overall. I cannot overstate how useful multiple shelfs are on an equaliser in a mastering context.

Most modern mastering EQ has a Q on the shelf. This is often called a Gerzon resonant shelf after the late Mr Michael Gerzon who proposed the concept. This comes in three forms that can be observed in figure 7.6: a half cut (C), half boost (B) or full Gerzon (A). The latter is the most useful in my opinion to accentuate the shelf which is the intent when applying a resonant shape.

This full Gerzon resonant shelf shape allows us to apply focus to the shelf lift or cut, but is especially useful when boosting and being able to create a highlight to the frequency corner. This has the effect of enhancing the lift and the amplitude shift is more active than it would be with a simple shelf. The focus makes the transition small and adds separation between areas of the mix. For example, a lift in the treble above the vocal at 8kHz can be achieved, while at the same time changing the impact of the vocals sibilance to reduce or boost, depending where the frequency corner is set.

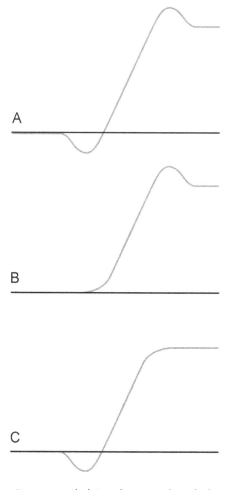

Figure 7.6 Shelving shapes and symbols

You can separate the bass end from the low mids/bass by creating a focused split around 400 to 200Hz. All of which would separate the lows from the aspects above to achieve more definition and often bounce in the music. This style of shelf Q is one of the reasons why mastering engineers like the Manley Massive Passive or a classic Pultec EQ can sculpt these types of outcomes. In digital EQ, most full feature plugins allow this outcome though often with a lot more flexibility with the type of Gerzon shape available having the option for an A,B,C types. This is not an option on analogue EQ where the type will be generally fixed.

High and low pass filters

To say a high pass filter (HPF) and low pass filter (LPF) should always be applied at the start of the mastering chain is misguided. If in analysis, there is a need because of a frequency issue, they should obviously be applied. But if a mix has been well managed, it will probably already have had filters applied on each channel to set the remit of each sound. The frequency balance should be well controlled in the extremities and already contained, creating a well-managed mix.

But, by applying a process, the context may change. For example, in applying a shelf either high or low, the extension of this shape would travel as far as the system resolution allows, i.e., very low (0Hz) and very high (70kHz). The context of those extremities has been changed, especially in the low frequency energy which more than likely will increase a noticeable effect on the overall RMS. Some of this will have little effect on the sound because those frequencies are lower than the audible spectrum. In this case, applying a HPF controls this aspect and focuses the shelf to a lift in the audible area of the mix. Typically, this would be set at 12–16Hz. In contrast a HS lift would excite the harmonic register all the way through the spectrum. This is good where required, but outside of this the boost can make our master sound harsh or crispy. The LPF smooths the response of the shelf to give a bell-like shape. This can be observed in figure 7.7.

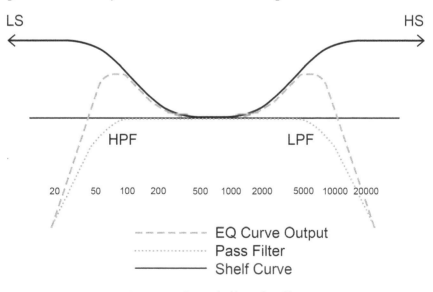

Figure 7.7 Every shelf needs a filter

Some engineers would use a bell instead to achieve the same outcome of rounding out the top end. With a more active sound to the EQ boost because it has a frequency focus to the bell shaped unlike the smoother HS/LPF combination. With some analogue EQ like a Maselec MEQ2, which does not have a filter set, this switch to a bell to inflate the top end without boosting the harmonic content to much is an elegant solution, especially as these bell shapes do not focus to resonance with higher gain. Practically I find using the bell as an 'Air' boost at the higher frequencies 20–30kHz with a wide Q of 1 often creates an effective lift to the top end without noticeably increasing the treble, but I would still apply a LPF above this to hem in the excess at 25–35kHz.

In another example, saturation might be applied to a mix by the tool type. This could be valve EQ or a compressor, or a soft clipper would again generate some harmonic outcome. The excess inflation of the very high end needs to be controlled as it can make the overall sound brittle or too hyped. Applying a low pass filter will correct this, but if set too low in the spectrum, it will remove audible musical content. To avoid this you can use upsampling.

Upsampling

To hem-in the top end effectively, our filter cannot just be set at 20kHz because it is the top of the audible spectrum and a filter does not start at the frequency point set. We define a filter frequency corner (the frequency set on the equaliser) as where the audio has been cut by -3dB. A filter set at 20kHz would actually start at noticeably lower at 14/15kHz at 12dB/oct, though at 6db/Oct it could be as low as 5/6kHz and this would graduate to -3dB at the corner frequency. All this means to control the harmonic energy from a shelf lift or saturation without cutting the musical part of the top end, our low pass filter frequency corner needs to be much higher than 20kHz, and 25–40kHz would be more appropriate to make the start of the filter touch the highest aspect of the audible spectrum at 19–20kHz. In analogue, this is not an issue because most high-quality circuits would have a very broad spectrum of 0–70kHz plus. But in digital at 44.1kHz, our usable spectrum would run up to 21.9kHz, 48kHz to 23,9kHz, but this is still restrictive as the filter would still clearly come into the music's main audible spectrum.

Those interested in the maths will quote Nyquist Theorem [6]. This never affected my choice of equipment, but the sound produced and the practicality in use of the equipment does every time. If you are interested in anti-aliasing filters and digital signal manipulation theory, there are many publications and arenas to explore this knowledge.

Coming back to the issues of practical available frequency range and filters. This is where upsampling in the digital domain comes into play; doubling the sample rate increases the available frequency range well above the normal available spectrum. 88.2kHz would go up to 43.9kHz, and at 96kHz reaches 47.9kHz. Upsampling a sound file does not make any addition to the audio, but any processing will now use this newly available resolution. Increasing the range gives the ability to set the correct filter corner frequency to hem in the shelf boost. Somewhere between 25k to 40k would normally be appropriate when applying a shelf lift, though the frequency corner set is dependent on the pass filter roll off. The softer the shape, the higher the frequency needs to be as a theme. This need to control the harmonic content of the high end is why classic mastering equalisers filter sets go much higher than 20kHz. For example, EMI TG12345 30kHz, Manley Massive Passive LPF 52, 40, 27kHz, Dangerous Bax EQ LPF 70 and 28kHz. All these have selectable LPF set well above 20kHz.

In practical use, to achieve a smooth lift to the top end and control the harmonic content, there always needs to be a LPF in conjunction with HS lift to appropriately control the high end. The same goes for the bass – an LS and an HPF control the extra amplitude that can become present in the LS lift. The HPF reduces inaudible bass energy and tightens the low-end lift. Looking again at classic EQ, the numbers observed are in the ranges of 12Hz/16Hz/18Hz to hem in the LS.

Upsampling in a general context increases the resolution of the equaliser when processing even when the original source is a lower rate. Processing at the higher resolution gives a more detailed response in the EQ shift even when downsampled again. This is why many modern digital equalisers upsample when processing the audio anyway. There are no negatives as long as the system manages this correctly. These potential issues are discussed in Chapter 10 'The transfer path'.

Working at these higher rates in the first place during recording and mixing means the production is already benefiting from the higher resolution in all regards.

Direct current offset filter

Direct current (DC) offset is where there is a bias to zero in an alternating current (AC) signal, where the audio crosses between the positive and neg-ative power as the waveform alternates. You may have heard the effect of this when making a cut in the waveform, and even though it appears to be at the zero-crossing point, it still creates a click. This is because the signal still travels to zero at the cut creating a vertical line in the waveform and an audible click. This would also be true of start/stop when an audible click will be present. This DC bias also changes the range of peaks on the positive or negative axis, depending on its positive or negative DC bias. A strong DC offset will clip sooner than if corrected. Fixing this is an uncomplicated out-come, a HPF at 2Hz or above. Remember, DC is just a horizontal line over time and its frequency would be 0Hz as it does not fluctuate up or down. Application of this filter removes this frequency range and effect. This is why some tools have an initial DC offset filter before gain stages and main filter sections to correct the signal right at the start of the chain, such as a TC electronic MD3 or 4. Most DAWs will have offline processing to correct DC. This can be simply applied before any post processing to the sound.

Filter types

When analysing mastering gear and the filters they use, different pole But-terworth and Bessel filters turn up more often than not, as opposed to mixing EQ where coincidental filters in dB per octave are the norm. Both Butter-worth and Bessel are linear in design and the group delay caused by the filter is equal to the roll off. This has the advantage of the filter removing less of the audio spectrum in the pass band from the corner frequency. Basically, they sound less intrusive across the spectrum, with smoother sound overall. The difference in these shapes can be observed in figure 7.8 in compar-ison to a classic mix console coincident filter. In practice this means the filter starts cutting much closer to the corner frequency, but still maintains a musical roll off that would be associated with a coincidental filter. This type of coincidental filter removes more of the sound which is to be avoided in a mastering context. When working on complex program material like a stereo mix, it is best to avoid removing aspects unnecessarily because com-pensating for this would be required using a bell or shelf to correct the loss.

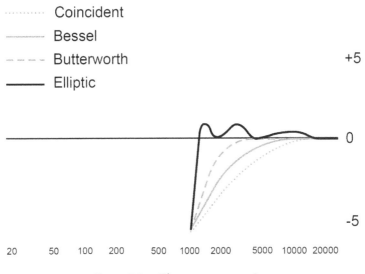

Figure 7.8 – Filter types comparison

It is always better to keep things as straightforward as possible, and use the right shape for the task at hand.

It is also worth mentioning an Elliptic filter which can be compared in figure 7.8. Most filters do not have a ripple up the spectrum from the frequency corner and are extremely smooth in roll off shape. An Elliptic or also known as Cauer filter will create a very aggressive bandstop, but also create a wave ripple up the spectrum in the band pass which will change the context of the tone. It might be tempting to think an oblique band stop at the bottom of the sub could give a tight control over the energy in the lows, but the negative effects of this are stark and sound very unnatural. It is a useful filter tool in restoration or cleaning up Foley work, but not for mastering in general in my opinion.

Summary

Before starting to use EQ in mastering, the principles of why a process works the way it does needs to be fully comprehended. This leads directly to correct tool selection in both 'tool type' circuit design and the practical requirements in a channel pathing perspective and step controls ranges. All this leads to a more effective manipulation of the frequency balance, assuming in our analysis it is considered appropriate to respond with a static outcome of EQ. If the evaluation links to a momentary response, the consideration needs to be focused on controlling the dynamics. To fully develop a comprehension of EQ tools, look to the manuals of those you have access to and EQ you aspire to own. Developing an appreciation of how EQ's differ will enhance your ability to make effective tool selection and get the most out of manipulating the tools you have access to.

Dynamic processors and tonal balance

Micro and macro dynamic

Dynamic observations can be broken into two distinct principles: the macro and the micro dynamic. Both these movements are happening in music most of the time.

The macro dynamic is when amplitude changes over a period of time, the crescendo and diminuendo of a piece of music, the volume lift between a verse and chorus, and so on. From a mastering perspective in the final outcome, this is one of the key observations to have balanced effectively. A consumer would not be expected to have to adjust the volume when listening to a song during its passage. An example would be where the dynamic range is very wide between musical sections of a song, this observation would be a macro dynamic issue. Listening to it in a car where there is a high background noise, the initial quiet level in the dynamic would mean the listener will have to turn up the volume to hear the music clearly. The amplitude would rise as the song progresses, and when the music reached the loudest sections it will likely be painfully loud as the macro dynamic difference is too wide. The listener would have no choice but to reduce the volume. This same observation may not initially be obvious when listening on a high-end system in a quiet environment because the scope to the dynamic range sounds impactful and emotive and no details are lost in the quieter sections. The mastering process must make sure the music translates effectively in all outcomes with minimal compromise. This type of macro issue should not be present in a commercial outcome destined for multiple playback scenarios.

The micro dynamic observation is the individual transients, the percussive parts of a song, the rhythm, but also the rest of that envelope, where

DOI: 10.4324/9781003329251-9

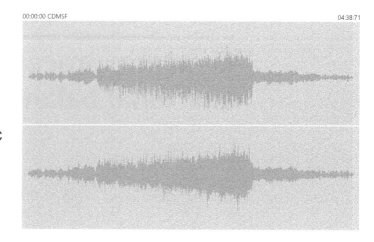

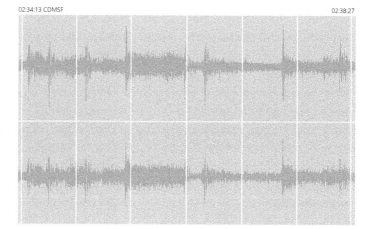

Figure 8.1 Micro and macro dynamic

the transient falls into the trough before the rise to the next transient hit. This envelope shape is the pulse of the music. Without appropriate control of this micro element, the music will lack life, pulse and forward motion.

Compression and expansion principles (upward and downward)

Compression means to reduce and expand to increase. In terms of dynamics this can happen in two opposing directions – upward or downward. In applying downward compression the dynamic range is reduced and reduction has been applied to the peaks and loudest parts of the mix (crescendos). This is what most people using compression are familiar with. But upward compression can also be achieved as an outcome, where the lower-level material in the RMS is increased at the trough in the micro dynamic of the waveform. Again, this is compression where the dynamic range has been reduced. But an upward compressor does not directly exist as a processor, it is instead an observable outcome. Upward compression is achieved by using a clean parallel path and downward compression. The summative effect of this if applied correctly achieves upward compression. A dynamics processor with a 'mix' or 'blend' control is still using a parallel circuit. It is not an upwards compressor but can achieve that as an outcome. It has the ability to be used as a parallel compressor. This principle in relationship between upward and downward can be observed in figure 8.2.

The same goes for expansion (see figure 8.2). To achieve upward expansion the dynamic range must increase by affecting the peak material and extending the transient. Downward expansion is what most engineers are familiar with, making the lower-level material quieter or off, a straightforward gating process. The dynamic range is increased as the floor level is decreased, thereby achieving a longer distance between the quietest parts of the envelope and the loudest peaks, thus expanded. Downward expansion would not be used in mastering as a norm, but there are always scenarios where this could be used, because there is something wrong with the source, i.e., requiring noise suppression in quieter sections. But upward

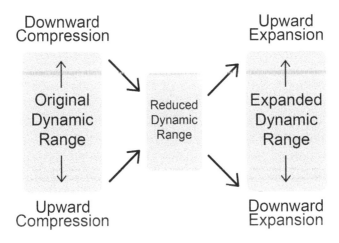

Figure 8.2 Compression and expansion principles

expansion can be an excellent process to achieve clarity in a master, especially to expand the Side elements to achieve clarity in the stereo impression, or it can be used across the whole stereo mix to uncompress an overly compressed source.

These principles of dynamic application are all available in a common mixing tool – the transient designer.

Positive attack - Upward expansion.
Positive sustain - Upward compression.
Negative attack - Downward compression.
Negative sustain - Downward expansion.

Two simple controls attack and sustain to deliver the four main dynamic themes. Obviously, this kind of restriction in control and automatic response does not lend itself to the discrete and finite control required in mastering, although it is a useful example in envelope shaping principles to effect tonal change on a given source.

Visualising the transient shape

It is helpful when developing our critical ear to appreciate the visual of the waveform shape and how it is changing with the dynamic's processor. Visualising what is happening to the waveform will help to comprehend the potential outcomes, not only in the gain change over time, but the tonal shift also achieved. The sharper the transient, the brighter the tone, increase the sustain, and more bass weight will be perceived.

During the section on ear training in Chapter 2, I discussed the principle of linking the observation of tone while listening to volume change in the audio. This could be in the micro or the macro. If the dynamics are controlled appropriately, this potential change over time in the perceived outcome should have been negated, especially in the macro, so the end user does not need to adjust the volume of the music during its passage. By changing the dynamic relationship in the music, there must also have been an effect on the tone.

Applying downward compression, changing the attack of the transient and reducing it is going to dull the audio because there is less attack in the sound. Sharp changes in waveforms relate to high frequency content.

In smoothing this energy there is less high frequency in the signal and hence it sounds duller. An extreme example of this would be applying a limiter where the peaks are cut off. When changing the shape of the transients, the tone has been affected, dulling the outcome. By the same principle, in achieving the outcome of upward compression, the trough in the envelope shape is increased in the RMS or sustain of the sound. The more sustain (release/decay) to the envelope, the more bass perceived relatively.

These outcomes are observed in the music's tone because of the application of dynamics. This has previously been referred to as the 'secondary outcome' of a process. The 'primary' is the change in dynamic range, the secondary is the change in tone. This outcome should always be compared at equal loudness or the effect of the change in amplitude will also be observed in the outcome. In manipulating the dynamic range, the following secondary outcomes will apply:

Downward compression > dulling.
Upward compression > increase in the bass.
Downward expansion > less bass.
Upward expansion > increase in treble.

The basic principle of tone and envelope shape are best observed initially in the adoption of the same visualisation that can be seen when using a synthesiser envelope generator – an ADSR (attack, decay, sustain, release). Utilise a synth with sine wave set as a trigger running into it ADSR. Repeat triggering a note and set the attack to the fastest, and the decay and release to the minimum. When adjusting the sustain between 0 and 100%, at zero the outcome would be a tick sound, treble, but when increasing the sustain control bass will be introduced as more sustain is achieved.

The same is observable with the voice. Creating a 'tock' sound with the tongue on the bottom of the mouth creates a short percussion sound which is perceived as trebly and percussive, with no low bass and some lower and higher mids. Singing a low sustained note, an 'oooo', the bass would be heard along with low mids and no attack and a lot less treble and higher mids.

It is the change in the shape of the envelope that delivers the change in outcome with tone. Fully comprehending this means that in applying dynamic change to a mix, control over the tonal balance of the song could be achieved without the need for EQ. Less processing to achieve the same outcome is always preferential in a mastering scenario. EQ is not always required to achieve an effective change in the tonal balance of a master if the source mix is positive, which in principle a good mix should be.

Manual compression

The simplest form of compression is 'manual', the adjustment of gain in the macro dynamic. Though not directly what would be observed as a processor, it is a process of amplitude adjustment, and a 'tool' to change the gain is needed to achieve it.

If a piece of music is levelled in amplitude across the quiet and loud sections by adjusting the fader, the overall range of the audio is reduced making it more compressed. Louder parts are reduced and the quietest sections are louder. Our perception of the tone at these points changes too, especially with the quiet sections being clearer, brighter and more forward. Yet the mix and its internal balance have not changed, just our perception of the audio relative in the context of the song. This is a powerful tool to achieve very positive outcomes with no actual change to the mix micro dynamic, tone or sound balance in the relative sections.

The best way to implement manual compression is not by automation, but with object-based editing with equal gain crossfades. This means utilising editing in the arrange/edit window of the DAW to change the amplitude of sections of the audio with each parts volume/clip gain while utilising crossfades to link. You can test the gain structure of your chosen DAWs crossfades editor using a sum/difference matrix. Simply make a copy of the original mix and polarity invert against the original and a null should be achieved. Create a crossfade in one of the objects, extend the cross fade out over at least ten seconds and play across the transition; nothing will be heard if the fades are equal gain. If you do observe audio in the difference, adjust the fades power until you achieve nulling. That is the correct setting as the starting point for all manual compression transitions. You can now adjust the split object to have different clip/object gain, and the crossfade will create a smooth transition between these sectional amplitudes. This principle can be observed in figure 8.3, the louder object at −3dB transitions via equal gain fades to −4.5dB to manually address a macro dynamic imbalance.

This object-based method also facilitates amplitude changes at the start of the transfer path, making all other processes post these macro gain changes. As a theme, this is the best place to make macro gain adjustments, directly

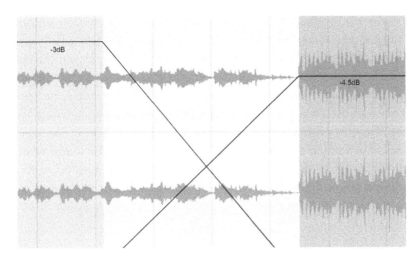

Figure 8.3 Manual compression achieved with object-based editing

to the original mix audio file, changing the observed macro issues first to reduce or negate these amplitude differences, especially in the tone. This approach pushes the corrected level directly into the start of the transfer path before any other processing has transpired. This makes the threshold of the following dynamic processors better targeted as the dynamic range is reduced to allow more effective control. It helps setting the knee because the scope is no longer overly large. Equally, the range sent into an equaliser is reduced, making the EQ'd variance less which creates a smoother resonance from the EQ overall.

Another advantage of addressing aspects on the object can be the transitions made, which are often in parts of the music that contrast. If any corrective targeted processing is required in those sections, it can be applied directly to the object. Not all DAWs allow for this in routing, but mastering orientated DAWs such as Sequoia or its sibling Samplitude do. You may have to render the object or split to separate tracks to achieve this otherwise.

On occasion, it could be applicable to apply manual compression during the sequencing process after a master print, maybe on an intro riff where the song just needs a little more push in the guitar level, or the tail out of the audio needs a lift. In principle, these would have been technically better applied in the pre-process. But practically when sequencing a whole album, it is acceptable to make these changes because they are often observed in the transition between songs rather than within the song in isolation.

It is also worth noting that manual expansion can be achieved by adjusting the same level, increasing the dynamic range across the macro. This would be infrequent because our concern is normally with achieving a more compressed/levelled outcome in mastering. Some mixes do need more contrast between the louder and quieter sections as the mix is overly compressed in the macro dynamic. If it is, a discussion with the mix engineer is a good option. The mix engineer might have left a bus compressor on the master. Just because a mix is supplied, it does not mean it should be mastered or is fit for mastering. It could also be in sequencing, a guitar intro needs more contrast, and turning the start riff down a little just gives the correct kick when the rest of the music comes in; this is manual expansion.

Full band or single band compression

Full band means the application of the dynamic process covers the entire frequency spectrum of the audio. This is important because there is no potential phase shift between the frequency range as there will be with a non-phase-aligned multiband processor. The input is the same as the output other than the manipulation of the amplitude and the latency the processing created. This is potentially very different to an equaliser where the change in the frequency band can induce phase distortion, an adjustment in time between frequency ranges. This is one of the fundamentals as to why dynamics are less invasive in processing than some EQ outcomes.

Single band compression is the same as full band, but the term single band is used when describing a multiband dynamics processor with all the crossover bands disabled. There is only one band covering the entire frequency range. An example of this would be the TC Electronic MD3 or MD4 where the number of crossover bands can be reduced until it is no longer multiband because all the frequency spectrum is manipulated by a single processing band.

Multiband or split band compression/expansion

Split band means there is one crossover, and only one side of that frequency range is being processed. An example of this would be de-esser, where the high frequency band is actively processing the audio. The lower band passes the signal with no processing. Or the active band is a bell shape with one frequency to set at its centre and a Q to adjust the width.

On the other hand, a multiband means more than one band could be active. Multiband compressors/expanders come in many different forms, but the basic principle is being able to target a given area of the frequency range to control the dynamic of the audio. Though caution must be used, because by splitting the audio into frequency sections and adjusting the amplitude inherently leads to the application of an equalisation process as well as any dynamic adjustment. It is critical to be clear about why that band is being targeted. A multiband compressor can easily ruin the mix balance of a song. Because of this amplitude movement of frequency range, it is changing the equalisation of the mix. As a theme, a multiband should only be used correctively to 'fix' something that is wrong in the mix and not to apply general compression.

When creating a frequency split in the audio spectrum and applying a process, in this case a dynamic one, the fact the bands overlap creates a phase difference. In the normal EQ process, this is referred to as phase distortion. In the digital realm, it can be negated by using linear phase filters and removing the phase distortion. This is why most digital multibands are linear phase or minimum phase, to focus the tool on correctively affecting the audio without additional artefacts. The principle of linear phase processing is explained in Chapter 7 'Equalisation and tonal balance'.

The other aspect to consider is the steepness of the slope in the crossover. The more aggressive the roll-off, the less the crossover is exposed and the more targeted the band, and the less natural the gain cut and boost as the band will become flatter at the top. This can sound very unnatural if not correctly targeted. A good example of this type of linear phase steep roll-off is the UAD Precision Multiband. A broader, more natural sounding crossover would be the TC Electronic MD3 which is very broad indeed. These broader bands are more useful for targeting compression on instruments as a whole, such as a vocal or the bass end/kick as a whole. The steep multibands are more effective at targeting frequencies, like resonance in the bass end or excessive snap on a snare drum or sibilance. The relationship between these different wide and narrow crossover shapes can be observed in figure 8.4. Pink noise has been played through each filter soloed with the crossover set the same at 315Hz and 3.15kHz. There is a flat response to act as a control for reference. The sharpness of the UAD is clear relative to the MD3. If you tested your multibands with a musical reference making the same A/B, you would be able to clearly hear the difference.

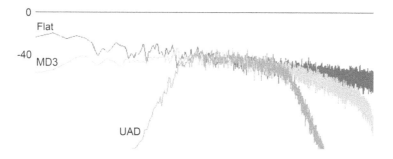

Figure 8.4 Linear phase multiband filter shapes comparison

Dynamic equalisation

An extension of multiband is where the bands operate as would be expected in a normal EQ shape, and not as in a multiband where the tops are flat. This inherently means they are more active, like a normal EQ with phase distortion present, such as the Oxford Dyn EQ. They often have a separate sidechain (SC) Q to focus the EQ's response. This is not always the case, as with the Weiss EQ1 where the SC is the area selected to EQ. As a theme, these dynamic EQs are more noticeable in use, but equally more musical as the response is a classic EQ shape. The other key factor is that the filter bands overlap, which is not possible with a multiband unless using multiple instances. This ability to make static and dynamic filter shapes work as one is a key aspect, I have found this very effective practically. I also find the expansion function on dynamic EQ more effective than a multiband expander as the shape is musical.

The Weiss EQ1 is especially good at this type of functionality where a few differing changes are required. This comparative between the shapes can be clearly observed in figure 8.5 where the multiband has a 'flat top' and the dynamic EQ has a classic 'bell' shape.

The sidechain/key input and the detection circuit

All dynamics tools control the reduction or expansion of the amplitude triggered via the SC bus. Even if the tool does not have an external SC input, it still has an SC that delivers audio to the detection circuit. The input to an SC is called the key input, but more often than not it is just labelled the SC. Mastering dynamics as a theme often have access to an external input to

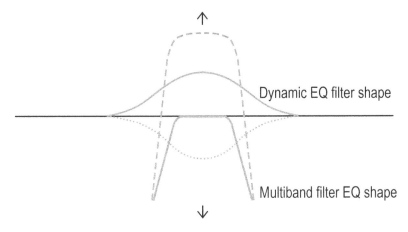

Dynamic EQ filter shape

Multiband filter EQ shape

Figure 8.5 Multiband dynamics filter shape compared to a dynamic EQ's filter shape

the SC bus, though some only have a switchable filter in the path such as a Manley Vari MU mastering version with the high pass filter (HPF) option. This unit's HPF is fixed at 100Hz, relieving the audio of the main weight of the bass and all the sub into the detection circuit by activating and deactivating the toggle switches. The main audio processing path still has the full frequency range of the signal, but the compressor is now being triggered without the emphasis of the kick and bass push. With heavier weighted music such as EDM or Reggae, this is often an effective way to avoid pumping, but equally with all outcomes, the lack of bass reduces the potential dynamic range and makes the compressor focus on the weightier elements left like vocal, solos or the body of the snare in the lower mids.

If the compressor has a discrete SC external input like a Weiss DS1mk3, the external input is just an audio bus that can be processed like any other channel. A filter or EQ can be applied to achieve the appropriate cut required. This could easily be set up as a trigger for just the sibilance, and the dynamic would be applied to the whole signal. This is more useful for a discrete vocal stem rather than a full stereo mix from a de-esser perspective, but comprehending the principle opens up many possibilities for processing.

All stereo dynamics should have the ability to link/unlink the internal SC (unless fixed as linked). When linked together it avoids the stereo image from being destabilised from the centre position. Without this link, the two channels operate independently of each other, making the amplitude responses difference in the L and R channels when the music is stereo. With

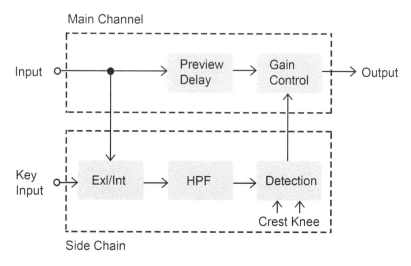

Figure 8.6 Dynamic processors signal paths

a mono source, the L and R response would be the same. This difference in amplitude between L and R will cause the centre of the mix image to swing around the middle position, like turning the pan control of a stereo mix one way and then the other.

With the dynamics processor signal split between audio process path and SC, the time can be adjusted between them. Delaying the process signal path makes the SC effectively ahead in time. This is called a 'look ahead' control. The function of the SC being an audio bus offers up a myriad of possibilities in how many outcomes of processing can be manipulated creatively. The SC bus is a very powerful aspect of dynamic control. The routing of these bus paths can be observed in figure 8.6.

Pathing 1/2, L/R, M/S

As discussed, when looking at what defines a mastering tool in Chapter 6, the principle of discrete dual mono means a process can access the individual matrixed parts of the mix. In conjunction with the external SC access, this means different channels of the matrixed mix can trigger one another. An example of this could be imposing the rhythm of the drums on the guitars in a rock mix. The guitars would generally be panned in the stereo field, hard panned, and they would appear in the 'Side' after encoding the stereo to MS. The kick and snare would be appearing in the 'mono' channel. The kick and snare would have a clear transient, so an effective trigger can be achieved from this in the Mid path. By linking the SC, the Side channel is also triggered reacting to the kick and snare to 'duck' the guitars and impose the main rhythm of the drums on the guitars. Simply unmasking the kick and snare away from the density of the sustain guitars.

As another example, cymbals are overly vibrant in the mix, but an effective trigger cannot be achieved in the stereo path because there are too many other transients in that frequency range (kick, snares, vocal). By observing the difference, the cymbals are clearly audible and more isolated. Encoding the stereo to MS, a trigger can be achieved from the Side, compressing the high frequency with a split band, but also with the SC linked, triggering the Mid with a differing or same frequency band to focus on the cymbal element in that part of the image. It is important to make sure the Mid channel threshold is very high to avoid any potential miss trigger from the Mid. As another example, triggering from the vocals in the mono can create a dip in the synths present in the Side to unmask the image and achieve a more compact dynamic range. The synths are wrapped around the vocal thus bringing the vocal forward in the spatial impression.

This use of LRMS with the SC is a powerful tool to achieve effective control over parts of the mix that would be impossible to access otherwise with corrective processing. This creates the glue for the music, which is the source of the triggers imposing a very musical response. Other parts of the mix context are affected with this same glue. This is not the same as setting these SC effects up within the mix or stems to correct. This type of cohesion is part of what mastering is all about.

Manual compression and expansion in MS

The simple action of decreasing or raising the Side in amplitude only is a form of manual compression or expansion because the relationship in gain structure is changing overall. There are a few rules to observe in this. One is not to collapse the matrix. The MS is using polarity and a summed gain to matrix the mono aspect from the Side. When changing the relationship of amplitude between the Mid and the Side, there is a possibility of collapsing the matrix. The most the relative level can move around the original position is plus/minus 3dB. This is a big relative gain change, but it can be easy to get carried away with width perspectives. It is important to have a context to view the stereo impression against. This is another example of where audio references are able to assist. It does not matter what the genre is, it is just the observation of the sonic width in comparison that is of interest. You should be able to clearly hear where the width sits effectively 'in sync' with the width of the reference. If this is a positive reference, it will have an effective stereo impression. The mono compatibility should also be checked after adjusting the MS amplitude balance to make sure the downmix is not being made to collapse in the mono or 'fold-in' because the Side element has been pushed too hard. This is especially important when a master is intended to be cut to a lacquer for eventual vinyl manufacture. Too much hype in the Side in relative gain will cause breakthrough during the cut. To avoid this, the cutting engineer would have to either reduce the width and/or reduce the volume of the cut leading to more surface noise relative.

Crest control

This is sometimes labelled as the peak/RMS or avg/peak and will be either a variable set of values or a toggle switch. The crest control is located in the detection circuit of the compressor fed by the SC. You can view this pathing in figure 8.6. This changes the focus of the threshold of the dynamics tool, in the same way as a standard peak/RMS meter would be observing the loudness in the RMS and the maximum amplitude with the peak measurement. The crest control enables this observation in the detection circuit of the dynamics tool. Nearly all compressors have a focus on peak intrinsic to their design, as the trigger is from the maximum amplitude (peak) as the norm. The ability to change the focus opens up the possibility of triggers from different aspects of the mix. Some compressors will have a straight switch, but others allow scope across the range of peak to RMS in dB. This would normally be a range of a general mix dynamic of approximately -24dB to 0dBFS. Therefore, sitting the crest at -12dB will allow a more equal trigger between peak and RMS. For example, this is effective to avoid transients when controlling a specific aspect in the mix. Vocals would be the most obvious example as they are loud in overall mix context and generally sustained. Hence the RMS crest setting allows for a cleaner trigger of this element in the complex programme material. This can be very effective to target aspects in the mix which are hard to achieve a response from with a conventional trigger focused on peak. As in the previous vocal example, the kicks and snare would trigger the compressor with a focus on peak, moving to RMS the sustained elements (vocals, guitars, synths) come to the fore.

Parallel processing

The basic premise of parallel processing is to apply a process and mix it with the original dry signal. In doing so, lessen or 'mask' the processing to reduce its audible effect, to make something that would be obvious and less noticeable.

One of the more helpful uses for this in the analogue domain is reducing the effect of EQ. Most mixing EQ has a sizable range of gain of 15/20dB. Most modern mastering EQ would be in .5db steps reaching a maximum of 5 to 8dB. Using parallel mix means these values can be reduced as the relative gain of the signal is 'mixing' in a 'dry' element. One issue that becomes apparent at this point is the phase difference between processing. One must be careful about how the paths are blended to avoid a phase difference. If significant, a delay would need to be created on the dry path to align.

This phase difference is more of an initial problem in digital systems because any calculation of DSP (processing) creates latency. Most modern DAWs will do this automatically using delay compensation, but some external processors (TC system 6000, for example) do not. Delay compensation should not be assumed and is part of our critical evaluation of the signal path. It is not uncommon for a DAW to run out of buffer/delay. Even though it says it is in time, it is not, again some DAW (Pro Tools) have a clear indicator when the delay compensation is not working. It is critical to test these aspects with your DAW to have a good understanding of how it operates.

To apply any parallel process, awareness of the signal path is critical to avoid introducing any unwanted artefacts. The easiest way to test any parallel path is by using a simple sum/difference matrix. With all aspects active, null should be achieved, or as close to null if using a processor without a fully clean tool type outcome. Again, sum/difference is a positive way to test whether it would be helpful to reduce or increase its effect in the context of the process. Parallel processes are covered in-depth in Chapter 12 'Practical approach to dynamic processing'.

Reference level/headroom control

On some digital tools, especially compressors, there is a reference level control. This is to change the relative perspective of the threshold where the input can be driven more or less depending on requirement. For example, if a reference level is set to 0dB and a threshold range is 24dB, the threshold can be reduced to -24dB, but we cannot go lower. If the reference level control is now reduced to -6dB, the threshold would now go down to -30dB, but the highest setting would be -6dB. The reference level enables a change to the perspective of the threshold across the audio. This is useful to be able to drive the input of the compressor to gain more saturation if the plugin is an analogue emulation. Sometimes this is referred to as a headroom control, but it still has the same effect, making the controls have useful context with the threshold range relative. In analogue, this outcome is simply achieved by sending more signal into the compressor to drive it; in digital this would potentially lead to digital clipping.

Why one will never be enough

To achieve the appropriate amount of dynamic control of a mix, it is highly unlikely you will be able to do this with just one dynamic process. As a theme, more than 2dB of continuous reduction with one process will be noticeable in its effect. To clearly hear the process, above 2dB would generally be required which may be the case with a soft clipper where the outcome wants to be heard, or a downward compressor pushing into the rhythm of the kick and snare. But this creative pumping aspect would generally have been created during the mix. In mastering, it could be enhanced, but again with a moderate gain reduction less than 2dB. Most dynamic processes in mastering are quite transparent to support the musical outcomes rather than impose a new perspective or creative effect. Three or four dynamic processes or more are going to be needed to achieve the kind of control required to compact the dynamic range to that of a modern master. Even when looking to achieve a wider dynamic range, the restriction will still be a notable amount. Focus with critical analysis should highlight where these restrictions and which types are required, making obvious where and how they should be applied in the transfer path. How to achieve this is explained in the rest of this book, assuming you have already taken onboard the ideas and learning from the previous chapters.

There is never too much compression, just bad compression

There has never been a truer statement in audio production. The number of processes is never the issue. It is the way they are used. Comprehending how the envelope is being shaped is key. Hearing the envelope of the original source means it is understood how to change it, and once applied, tell how it has been shaped for the better. A sound or mix is compressed enough when it has had enough dynamic control. This is often when the instrument holds position in the sound balance depth of field. It does not fluctuate in perspective meaning it is compressed/restricted enough. This is simpler to control in the mix than the mastering; positive sound balance inherently leads to a more effective and detailed mix, leading to a more effective master.

This compressed and controlled sound balance is in part why it can be perceived as easy to mix programmed sounds as they are by their nature repetitive and compressed. Inherently delivering an even (finished) sound from the outset. This does not mean the music created has anything of interest about it or even the mix is sonically positive. The same could be argued with a rock band. Give an average band an amazing drummer to deliver a concise compressed performance in recording and it will immediately sound more finished. This does not mean the song is any good, or the other performances are up to scratch. The sonics are only a part of the production chain from the concept through to performance, as explored back in the 'chain of production' discussed in Chapter 1. But consumers are used to hearing compressed music in the majority of genres. Making compressed sounding presentations appear more professional or commercial in outcome. This does not mean they are finished, just compressed sounding.

There are many aspects about dynamics to explore, such as controls, processor types, methodology, parallel processing, loudness, saturation and limiting. All this and much more is discussed in a practical context in Chapter 11 'Practical approach to dynamic controls' and Chapter 12 'Practical approach to dynamic processing'.

Contextual processing analysis

With constructive, comparative and corrective analysis complete, there is the need to divide our observation into two clear groups: 'corrective', where mix errors have been assessed, and 'overview', where the music's general perspective is being examined. Both aspects have been observed in analysis simultaneously, though it is important to make this clear distinction for the benefit of our eventual approach to processing.

Quantitative evaluation

The method of linking analysis to a process can be broken into clear stages: those observed as potential tonal and dynamic processing in overview of the whole sound of the mix, and corrective processes that involve a potential change because a mix aspect is wrong. With an excellent mix there would be no corrective processing required. This points to the need for corrective processing to always take place before any other aspect evaluated as processing in overview. Breaking our observation down to make clear if the analysis was about an individual instrument (corrective) or forming part of the general mix/musical context (overview). This is the clear distinction between something in focus, such as sibilant vocal (corrective), or micro dynamic variance with the snare (corrective). Or the whole mix sounds too dynamic in the micro (overview) or just lacks low end (overview). Once our observations have been established in these terms, it is equally important to clarify whether each assessment is momentary or static.

DOI: 10.4324/9781003329251-10

Momentary or static

To comprehend the context of potential tonal issues observed in analysis, it is helpful to split our assessment into two categories. Momentary is where the sonic in question increases and decreases in amplitude in the micro or macro dynamic, such as an uncontrolled snare or vocal could. Static is where the sonic in tone is present constantly throughout the song, such as a bright mix. This is not to be confused with a momentary bright mix where the treble would be overly dynamic in the micro dynamic of the source. An example would be where a constant hi-hat delivered to much hi-end, but the rest of the mix was relatively dull.

From a corrective perspective, to 'fix' an aspect in the mix, vocal sibilance, or bass note resonance. The target is only happening at momentary points in time during the passage of the song. Therefore, it should be targeted in the frequency range assessed but dynamically. The processing outcome only affects those points in time when the tool needs to be active. Making the approach very targeted as opposed to a static process which would affect the sound 100% of the time throughout the track. In contrast, when the low mids are overly weighted at 300–500Hz constantly during the passage of the song, you have a static outcome, and a fixed process in overview would be an appropriate response. As a theme the majority of momentary outcomes tend to be observed on processing that would have been associated with corrective processing. But there is always an exception that would fall into overview.

Qualitative evaluation

After ordering observations as corrective and overview while evaluating each as momentary or static, each observation should then be evaluated and linked with a process to change the outcome. Be careful at this stage not to jump ahead and link directly to an actual tool. As an example, when a track is too dynamic in the micro in overview, a link could be made to downward compression. Do not jump to link this to a processor such as a Shadow Hills Mastering Compressor or Weiss DS1-MK3 and so on. Each of these compressors will apply downward compression, but also have a potential tool type outcome from the circuit type. It is best to make those positive links to tool type later in the evaluative process when a clear map of the whole processing context has been made.

This also keeps our focus around the principles of processing and processors, and not become reliant on any one piece of equipment. The skill is in the knowledge of how to change an observed aspect of any sound whether simple or complex. It is not about what gear you own or have access to, but the ability to apply that knowledge effectively. The equipment helps, especially the listening environment and monitor path, but only when you actually understand how to get the most out of it.

Observation to processing: Contextualising the links

Visualising the difference in the frequency balance to our references, draw on the audio reference to hear the contrast in tonal shape. Observing these differences as momentary or static and dividing these assessments into corrective or overview makes the link to potential processing tools required more explicit. This visual method of observing the flat 'ideal' of the audio reference to our mix in analysis can be observed in figure 9.1.

A momentary observation in principle would link to a dynamics tool to correct. But if the aspect was static, an EQ filter shape is the right response. If the observation from analysis was that the mix was dull, this static observation would link to a HS to adjust. Observing too much power in the lows, the link would be to an LS cut or better to apply an HS treble boost. I cover the principle of when to cut and when to boost in Chapter 13 'Practical approach to equalisation'. These shelf responses to a static observation are opposite to an

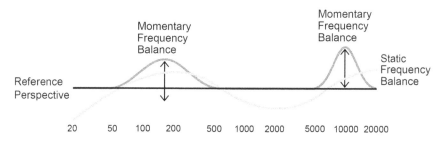

Figure 9.1 Momentary/static differences in tone between a 'flat' reference and mix source.

aspect in the top end observed as momentary, and the link would be to a multiband HS, or split band compressor focused on the high end. The shelf shape is maintained, but controls the aspect dynamically to smooth the response of the dynamic range to hear the actual tone in that area of the music. The more critical our listening becomes, the easier it is to hear the mixes true tone presented between the overly dynamic aspect during initial analysis.

In terms of full band dynamics, we look back at the principles to affect dynamic change. The three mainstays discussed were upward/downward compression and upward expansion for making clear links to our analysis. For example, if the observation was overly dynamic in the micro, the link would be downward compression. If observing masking or lack of presence in the quieter sections of the music, the link would be to upward compression. If the mix lacks life or bounce, the link would be to upward expansion. When making these links, it is always helpful to note down potential frequency ranges observed and key dynamic settings in approach. This is all part of developing your critical listening skills. In application you will find out how close you were or were not to these notes to make the assessment easier the next time. Link listening to the numbers. It can also help to observe where frequency ranges are grouped or overlap to help avoid a potential process working in counteraction with another process.

Below is a set of examples contextualising the observation made in analysis to a potential process principle. This is sectioned into observation in overview and those that would be corrective in focus.

Corrective processing:

Vocal is too sibilant > Split band compression HS 6kHz
Snare attack is aggressive/too sharp > Multiband compression bell 4–6kHz
Kick lacks attack > Dynamic EQ expansion bell 60–80Hz
Main vocals overly dynamic > Multiband compression bell 300Hz-3kHz
Guitar has too much snap > Dynamic EQ bell cut 3kHz Q1

Processing in overview:

Lacks sub > LS boost 35Hz with HPF 12Hz
Too dynamic in the macro > downward compression (slow attack)
Dull > HS 3kHz, LPF 30kHz
Lacks width > S+

Too much bass > Bass LS 250Hz cut + HPF 18Hz and/or HS lift + LPF

Macro dynamic is large, start section is quite relative > Manual compression

Too much low mids > Bell cut 400Hz or LS 200Hz, HS 600Hz boosts + HPF/LPF

Micro dynamic is large/sounds thin > Parallel compression

Aggressive random transients > Limiting

Further examples of this contextualisation of analysis linking to tools can be found in the online resources supporting this book. Look at the mastering analysis worksheet examples online and read the chapter 'Bring it all together: Engaging in mastering' for context. The link to online assets can be found in the introduction. Better still, digest all the learning to be gained from the book chapter by chapter, it is sequential.

Linking process to pathing: LRMS analysis

We now have a list of observations, each with a link to a potential process to positively affect it. LRMS analysis needed to observe where each of these processes should sit in relation to the audio path, i.e., channel 1/2 L/R M/S or stereo. For example, if a vocal is in the mono aspect of a mix and our analysis noted the vocal was too dynamic, the link should be to a multiband in a bell shape to correct. Its pathing would be on the M only, channel 1.

Main vocals overly dynamic > Multiband compression bell 300Hz-3kHz > Path M

At this point in the contextualisation of the process, it is helpful to be conscious of the potential for the dynamics processor's SC and the manipulation of the detection circuit of a dynamics tool. In context of the previous example, the vocal which is pathed in the mono aspect of the mix should process on the M, but as the compressor has two channels, it will also have the option to link the SC. This SC should be unlinked because the trigger is in the M, but if the backing vocals were in the S element, linking the SC would achieve an effective trigger from the M to push the S down at the same time. This reduces the backing vocals in tandem with the main vocal dynamics, maintaining sound balance between both while correcting the overly dynamic main vocal. Incorrect triggers from the S can be avoided by taking its threshold up above any trigger point. This is only possible with the controls unganged. Without the SC linked, the backing vocals are likely to come forward in the mix balance because the main vocal will be more controlled in dynamic range. Determining which context is correct comes from comprehending your analysis of the source mix correctly. Are the backing vocals masked by the main vocal? If so, unlink SC. Are the backing vocals correct in balance to the main vocal? If so, link the SC. It is all about contextualising analysis directly to the tools used to control the outcome in a way that improves the aesthetic in all regards.

Consideration should also be given to the crest control in the detection circuit of a dynamics tool. Continuing with the vocal example, the music will have transients from the drums/percussion, though having set crossovers around the vocal at maybe 300Hz-3kHz to target the main notes, there will still be a substantial transient from the snare and maybe the knock of the kick if it has a strong beater sound. Triggering from the peak (as most dynamics do by default), the compressor will react to the transient as opposed to the signal to be accessed, the vocal. This part of the signal will be sustained in the envelope, mainly containing RMS energy and little peak. By setting the crest control to RMS, with the SC trigger dedicated to M and crossovers applied, it is likely that a very effective trigger to control the overly dynamic vocal will be achieved, by pulling the louder aspects down and levelling the vocal perspective in the mix balance. All this would be noted as in the example below.

Main vocals overly dynamic > Multiband compression bell 300Hz-3kHz >
    Path M > Crest RMS / SC off

This gives a clear context to approach the setup of the processor and importantly why it is being applied. If this does not work and a trigger cannot be achieved, our initial analysis is probably flawed and it would be best to go back to critically re-evaluate the mix with this new context. Maybe it is not the vocal that is overly dynamic but the whole of the mix and it simply requires a full band compressor to control the apparent micro dynamic aspect from the vocal. Evaluating pathing, maybe the M would have the best trigger with SC linked to maintain mix image balance, and an HP SC filter would stop the compressor from pumping on the sub/kick. Riding the mix level overall as one is often referred to as mastering glue, even when applying an SC filtered target to the vocal and achieving the desired outcome in control. This still glues the range used together as other parts change because of this process triggered by the music itself. This is very different from placing the same process on the vocal stem as none of this interplay would exist.

Tool type

A clear list of the processing intent permits the linking of each process selected to a 'tool type'. When the full context of each process is known, it is possible to contextualise each in turn and ask ourselves what effect the process is meant to have on the sound. With corrective processing in general, a link to a clean tool type or a linear phase process would be appropriate as the audio should be corrected clinically. For example, our corrective main vocal example would be noted as.

Main vocals overly dynamic > Multiband compression bell 300Hz-3kHz > Path M > Crest RMS / SC off > Linear phase processor.

Potential tool type outcomes adding phase distortion or saturation/harmonics are not needed with processes to 'fix' the mix. In contrast, with a shelf boost to correct a dull sounding track in overview, it makes sense to add a processor that creates excitement that has an active circuit type such as a valve processor. This added harmonic content is enhancing the audio in application, and less of a shelf boost is required to achieve the correct lift. If less impact from the tool type was preferential and a clean tool type is too passive, the selection should be for an analogue style EQ that still introduces phase distortion in the boost adding some enhancement/character. But by applying this tool type and because it is a shelf, it would also be appropriate to add an LPF in the upper range of the spectrum to hold in the potential harmonic/excess air generated by the circuit type and the EQ shelf boost itself. The latter is discussed across the chapters around practical approaches to the application of processing.

Transfer path construction

Having established the potential processing, evaluating how all these processes affect each other is required to achieve the optimum outcome from our selected tools relative to analysis. We already considered 'corrective' and 'overview' in terms of path. It is now necessary to optimise the processing order to achieve the simplest path preventing duplicate processing and achieving the most cohesive order to avoid one process working against another.

One of the principles to observe is the effect of pathing order on the dynamic's tools threshold. When removing aspects of the frequency range, especially in the longer wavelengths, the relative trigger level range for the following processes is reduced. For example, if the vocal was overly dynamic, it should have been noted that this could be fixed with a multiband bell. Once controlled, any following overall downward compressor would be triggered from this corrected smoother outcome. In contrast, applying the overall downward process first would make it more reactive to the vocal level. Following this with the vocal multiband takes this dynamic aspect away making the manipulation of the mix disjointed in comparison to applying the corrective multiband first. It will also be more difficult to achieve an effective trigger from the corrective multiband as the overall downward compressor has controlled some of the vocal dynamic.

That is why splitting processing into corrective and overview in the previous tasks is important. Other processing will affect the threshold such as a static EQ change in the low mids or bass. If this is a cut, power is being removed that is not required. Hence this should go before the dynamic processes. As a theme, processes that remove energy should go before those that have an addition or add cohesion. For example, if during analysis there was the need to remove the lower weight from the mix, a HPF and LS cut overall could be used. This loss of energy will affect the dynamic tool's threshold whether the crest control was set to peak or RMS. The HPF and LS should be in front of those other processes in the chain that would be affected by it. Even though the process was observed as being in overview and not corrective, the focus should be on avoiding processing triggered by a signal that is taken away later in the chain. If our routing order was limited, as sometimes it is when working in analogue, an HS and LPF in addition to an HPF to control the very low sub would affect a positive addition to negate the impression of too much bass by increasing the highs – thinking positively with EQ. This approach is discussed in Chapter 13 'Practical approach to equalisation'. In this case the compression could now come after the EQ as both processes are working together. As there are many paths from input to output, it is important to consider all the consequences to achieve the best musical effect.

In another example, when applying HPF and/or LPF where these pass filters are going to be placed in the chain should be considered. If the LPF was selected because an HS was required from analysis of the mix, the LPF is there to hem in any excessive 'air' created from gain lift. When cutting the frequency range, it could be argued the cut needs to be at the start of the chain because energy is being removed and all post processes are affected by the change. But because the LPF is there from its association with the shelf, it makes sense to keep it after the process causing the change.

In considering the tool type, if the EQ creating the boosts is a valve circuit type, it will generate harmonics and any excess should be controlled. Though if an LPF has already been applied at the start of the chain, this filter addresses the unwanted frequency potential in the mix at the start of the transfer path, and another LPF should be applied after the valve processor albeit likely higher in the frequency spectrum. Overall that leaves two LPFs in the chain. Do not be afraid of adding the same process twice if there is a rationale for it.

Where does the EQ go?

Dynamic range is the key to this thought process. If the source mix is very dynamic and there is also a need to change the tonal balance. There is bound to be an element of the tone heard that is momentary, meaning loudness and its effect on tone must be at play. But if the observation is consistent over time, a static process of EQ would still be appropriate. The question is where to put the EQ in relation to the dynamic processing needed to control the overall dynamic range.

When applying an EQ boost, the micro dynamic aspects (loudest transients) will proportionally have more EQ applied than the quieter parts because they are louder. When compressing the signal, this difference in EQ application is imprinted into the audio even though the more managed dynamic range is controlled and the tone from amplitude difference is mitigated. The change in tone on the source from the EQ can still be heard in the compressed signal. It retains some of the impact of the dynamic of the original, but now with a more compressed outcome.

In contrast, compressing the signal first, the dynamic is controlled and the movement across the audio is more even, making the tonal shift from amplitude mitigated. When applying the EQ boost post, the tonal change is more proportionally distributed across a more controlled dynamic range. This outcome sounds softer, less EQ'd, but equally less dynamic in perception even though the two dynamic ranges will be very similar. Neither scenario is necessarily more correct than the other (as both can actually be correct). It is down to the aesthetic you want to achieve with the given audio.

Considering an EQ cut, in removing frequency range that is not required, logic would suggest cutting first, and process any dynamics after. It makes no sense to process a sound with dynamics and then take away part of the signal it has been responding to. The outcome will be disjointed, hearing dynamic control of the amplitude based on sound that cannot be heard. The caveat to this would be when controlling the music's dynamic range to a large degree; the secondary outcome effect of the dynamic processing would mean a cut may be needed to rebalance the frequency range. This cut would be post as the mix in general has been brought under control at this point in the path. This principle is discussed in Chapter 11 'Practical approach to dynamic controls'.

But as a theme, removing first and boosting last will always work well. If you want to achieve a more active master, try boosting first, pushing into the dynamic processing. There is also a frequency range density element to consider. If you have a mix with a weak sound balance to start with, you are likely to want to increase the lows. As with the argument for cutting first, processing the outcome that is more balanced is generally the best option. Logic would dictate boosting the lows to make the track tonally balanced before processing the dynamics to achieve the basis of the desired sound. I would tend to opt to change the frequency range before any other processing of the music if it has a noticeable imbalance, even if this has been observed in overview. The better the mix, the less this thought process is required and focus can be placed on the details with the dynamics to make the music move.

Upward before downward?

When considering restricting the dynamic range, it is likely there would be a need to change the perspective from both directions if the range is wide. Processing in just one direction, the effect would be heard more so than if it is balanced across the whole dynamic range. It is often normal to use a combination of dynamic processes. So, what should go first, upward or downward compression? Considering the effect of both processes, as a theme downward is reducing the power of the louder elements in both the micro and macro dynamic. Whether the crest trigger is from the RMS or peak, the outcome is overall gain reduction in the dynamic range from the top down. But considering upward compression, a parallel path is needed. The compressed part will be triggered from the peak information. The contrast in how much reduction is applied comes from the distance in the peaks to the RMS, the loud to the quiet, thereby achieving reduction to make the quietest aspects have no gain reduction and the loudest the most. It is then the sum back with the original signal that achieves the effect of upwards compression. This method is called parallel compression and is discussed in-depth in Chapter 12 'Practical approach to dynamic processing'. By this observation, you want the largest dynamic range to work with as the source for the upward compression to achieve the most effective result. As a theme, upward first, downward after. If the audio is really transient, it would be positive to consider limiting or soft clipping before the main dynamics processes to restrict the maximum range, thus achieving a more controlled ratio to the max threshold with the upward outcome, controlling in effect the range of the knee from threshold trigger to maximum. Practically this means a bigger ratio can be applied while still avoiding potential pumping.

Practicality of processing

When working purely in the analogue domain, there is a level of restriction imposed on the pathing because of the physical limitations of the equipment and routing available. Whereas, in the digital domain, there are infinite possibilities to the routing. But as the phrase goes, 'less is more', I feel this always applies to our audio pathing. The fewer pieces of gear in the way of the audio signal the better. For example, having assessed an EQ pre and post a dynamics process would be appropriate, by the restrictions of the analogue path the option would be to opt to use the EQ either pre or post to achieve the outcome. If the EQ involved cuts, this probably would be best pre by the logic discussed previously.

There may be potential restrictions on where the parallel circuit is positioned on our mastering console, or equally the main downward compressor. This inevitably leads to compromise, but getting the benefits out of the analogue path, the process is going to have a bigger effect than the pathing – something to always remember.

This simplification of the transfer path can have the added benefit of narrowing our options to hone our decisions directly back to the source of the original analysis, as that is why the tools are being applied in the first place. It is easy to get lost in the gear, micro tweaking and forgetting the bigger picture of enhancing the quality of the music overall. I would always try to simplify a transfer path where I can to achieve focus in processing relative to this bigger picture.

Summary

Remember, there is not a correct answer. You need to judge the best options for pathing based on the tools you have. The tool type affects the outcome, probably as much as the path routing will. Equally the order will have a smaller effect than the tool type selection, and none of this will have as much effect overall as the main primary processes. The more all these elements can be refined to work together, the better the outcomes will be. More important than all of the gear and pathing is the analysis; without that, you are just moving sound around with little purpose. Critical analysis with positive contextualisation in linking to tools, the type and LRMS pathing will make the audio sound significantly better and more musical whatever the genre.

The transfer path

It is important to consider the journey the audio takes through our processing. In mastering this is referred to as the 'transfer path', the carrier of the audio signal. In the early days of what would become the mastering engineer's role, the transfer engineer took the audio from recorded tape and transferred it by cutting to a lacquer. When transfer engineers started to enhance this transition with processing between, the mastering engineer was born.

With the signal path purely in the digital realm, this would seem uncomplex in principle, but even in this path there can be issues that need to be addressed to avoid changing the sound unintentionally. In analogue, the audio path quality is obviously critical, but equally in moving between realms, the transition in converting between domains requires scrutiny.

Analogue versus digital

There is no 'versus' in my mind; analogue is excellent for some sonic changes, as digital processing is for others. The skill is in making the correct choice based on the outcomes and parameters required for that individual master from analysis. The consideration from a transfer path perspective is in the consequence of moving the audio from digital to analogue and back again to digital. Without getting deep into the principle of digital theory, our main focus has to be with the quality of the conversion and how to maintain correct clocking throughout the systems used.

An incorrect timing with a digital audio clock is referred to as jitter because the samples are making irregular movements, unsteady, and

DOI: 10.4324/9781003329251-11

jittering around. Jitter within a system is not necessarily permanent; a 'bad' clocking outcome during playback does not affect the source audio. This is called interface jitter, but if recorded, the jitter error is imprinted into the audio. This cannot be undone as the original clock was in error and the correct clocking is not known. If it was there would be no jitter, as it would have been perfect in the first place.

This recorded clocking error is referred to as sample jitter. For example, in playing a mix the interface jitter may be affecting what is heard on the monitor DAC. But the audio is not actually affected in source, it is the playback that has the clock errors. Bouncing the audio offline, the DSP would process the signal creating samples one after another to assemble the output file. The clock is not used in the same way as when working in real time where the playback is at a fixed rate/interval. When processing offline, the jitter does not exist in this same context. But in playing this rendered audio, the jitter could again be present in the playback if there is a bad clock internally and interface jitter would again be present. By bouncing this audio in real-time by creating a digital loop back in the system, the interface jitter would be recorded back in with sample jitter recorded in-place.

When sending this signal through analogue gear and recording back in, even onto a different system with a perfect clock, the interface jitter is in the playback system and would now be recorded in the new system as sample jitter, because it is imprinted in the waveform as with the digital loop back. These small errors in the clock lead to the waveform being misshapen as the time errors move. This reshaping the waveform position can also be referred to as distortion because the waveform is distorted in shape. This does not sound like a fuzz box has been applied, but it does change the listening perspective, often thinning the sound overall as harmonics could have been created.

Once embedded, there is no way to know what the correct clocking was, it cannot be undone or de-jittered. In a lot of modern mastering setups, analogue processing is used as an insert into the main digital path. The playback and record systems DAC/ADC are one and the same in terms of clocking. If jitter exists, it is going to detrimentally affect the audio outcomes. This leads to the simple observation that the better the system clock and correct linking of the clocking between equipment, there will be less chance of any jitter. Ideally there should be none at any point in the whole system. Mastering is about identifying and implementing the ideal.

There are some easy rules to help avoid jitter. When recording, always use the clock on the ADC as the main clock. It is the piece of equipment subdividing the audio, and it is best to let it make those clocking decisions. In playback, set the main system interface as the clock source, this is internal on many systems. If your ADC and DAC are in the same unit, this may not be possible as the unit will have one main clock setting for all inputs/outputs (I/O).

Our monitor DAC, if separate, will resample the audio coming in buffering and playback with its own internal clock to negate any potential interface jitter. If your monitor DAC is part of the main converter, again it will be part of the internal clock. This is why many mastering engineers prefer the monitor DAC to be separate to the transfer path DAC and ADC used as an analogue insert in the digital path.

One way to maintain a quality system sync is using word clock. But you would note a dedicated standalone monitor DAC does not always have a word clock input. This is because digital input rejects interface jitter by buffering the incoming data and resyncing using its own internal clock to achieve an accurate reproduction. It ignores any clocking data and resolves the clocking speed based on the detected samplerate. This is referred to as asynchronous clocking because the actual clock in the DAC is not locked to the main system clock. While an AES/EBU interface connection [7] contains system clocking information, some system devices such as external digital hardware have separate word clock input to maintain a synchronous clock. This does not mean the samples are at the same place in time as any DSP has latency. For example a TC electronic System 6000 with the correct interface board installed can lock to word clock to maintain concurrent clock speed but has a latency from input to output. A Weiss EQ1 operates on AES/EBU digital connectivity synchronised to the selected system clock delivered via the AES/EBU to keep everything in sync, but it still has latency. You could use either of these devices in analogue with dedicated standalone conversion using their own internal clock. Again, it is a preference for some engineers though it will still have latency relative to the use of its DSP.

If you are interested in the protocols and principles behind digital interfacing, the Audio Engineering Society (AES) library is an excellent place to start, look at AES3 [7] standard as a starting point. The future of general digital interfacing is currently Dante, AES67 [8], though careful consideration of the clocking source is needed from a mastering perspective.

Word clock enables a unit or dedicated clock device to become the master for all other digital devices to maintain sample accurate synchronisation. This is very helpful when using multiple digital devices in any system setup, more so in recording or mixing where multiple interfaces need to be 'in sync' between each other to play and record all parts with the same samples accurately, making sure all the recorded parts are in-time with each other. Otherwise the drum overheads and room microphones recorded on one interface will be out of phase with the rest of the kit recorded on another set of interface inputs. In mastering, the transfer path is sequential apart from any parallel bussing, hence latency or sync in this way is not an issue from start to end of the path. But the accuracy of each part of the system's clocking is critical to avoid distortion of the signal through jitter. There are companies like MUTEC that make a dedicated Audio-Re-Clocker to remove jitter in the interface between digital hardware.

In summary, clock to the ADC when recording in, lock to the internal clock when playing back and use word clock where available – three simple principles to keep you in-sync. If you have a dedicated standalone clock unit like an Antelope Audio Master Clock, this would be the master in all scenarios syncing everything via word clock and/or AES/EBU, though the independent DAC would still be asynchronous.

Upsampling

It is still the case that many projects are recorded and mixed at standard sample rates of 44.1kHz or 48kHz. There are several rationales for why, but the main factors are available DSP and interfacing channel count. Both are often halved when doubling the sample rate. Given that recording generally requires a high track count in the first instance, and mixing equally requires a lot of DSP, it is understandable that the project has been created at those standard rates. In principle working at the highest resolution possible is ideal, but reality can dictate a lesser outcome.

In receiving mixes at these standard rates, there is a choice to either run with the sample rate as supplied or upsample to a higher rate. Upsampling does not make the source sound better because it is only the resolution carrying the audio that is changing and the original format is still inside this higher rate. This assumes the sample rate convertor is set to the highest quality possible. Historically sample rate conversion was DSP hungry. To save time when making test outcomes, the conversion quality could be reduced at the same time. In a modern context there is no need for anything other than the best all the time. You need to check your DAW preferences to fully comprehend what happens to the audio in conversion in differing scenarios.

Our upsampled audio is just the original rate in a new carrier that benefits when processing starts. All the calculations applied are now at a higher rate and more accurate than at the original rate. In the same way a higher bit depth increases the accuracy of processing as more dynamic range is available. This is especially beneficial when summing. In upsampling if the path is going to analogue and back digital, the benefit of capturing the analogue return at the higher resolution is also gained, capturing more of the analogue domain. In the digital realm, the manipulation of EQ is clearly enhanced in accuracy.

There is another solution to this if using two separate systems – playback at the original sample rate and capture from analogue at a higher resolution without any clocking link between the systems. Equally this principle can be used to downsample without the use of any sample rate conversion. Some engineers prefer this method of a separate playback and capture device so they have independent control of clocking and resolution in general.

It is also interesting to note that high definition (HD) formats require higher sample rates in the final master, 96kHz, 24bit being the current norm. Having upsampled the path at the start of the chain, the audio outcome produced through processing will have benefited from the higher resolution. A true HD mastered outcome has been created even if the original mix resolution was lower. Remember, a HD format cannot ever be created by simply increasing the sample rate and/or bit depth of a file. Actual processing change has to have been undertaken to extend the bit depth and sample rate to make it valid as part of the new delivery of the sonic.

Transfer path considerations

Absolute transparency in the transfer path is crucial. The tools placed in it should be the portal to the sonic changes required. Even in analogue this premise applies. Our mastering console may have the ability to drive to saturate the signal, but if this was not required, it would be transparent at an ideal signal level – unity gain, our reference level. Think back to our discussion about metering and calibration in Chapter 2 'Enabling analysis' and Chapter 3 'The monitor path'. I have been fortunate to have had the opportunity to work with several mastering consoles in my time, as well as a lot of the high-end mastering compressors and equalisers in analogue. They do truly shine when the path they are ported through is pristine. This gives the widest possible scope to outcomes and potential stylistic choice for differing genres.

It is also important to note the separation of the transfer part of the mastering console from the monitor path. Some consoles contain both, but their paths will be discrete. It often facilitates easing monitoring of the input (I/P) and output (O/P) of the transfer path within the monitor section. This would generally also have a dedicated metering output to observe the level of these differing points in the signal path. In principle it should be kept level to maintain equal loudness and maintain quality throughout the system. Keep the ideal level around the reference level (unity gain) if the system has been effectively calibrated.

In the modern construct, more and more digital pathing and processing is being used to achieve high quality outcomes. This does not mean that I think analogue processing is dead, far from it, but the nature of recall and revision in the modern construct forces the ability to revise and replicate a given signal path with ease. Some manufacturers have responded to this in analogue hardware with digital control and interfacing such as the units from Bettermaker, but others have gone down the route of potentially modelling their units in digital with better and less effective results. But the march of technology will bring improvement, as is obvious even when looking back at the last few years.

Metering

To effectively control the dynamic range of a system, it is important to be able to correctly observe the signal at varied potential points of clip in the transfer path, and also be fully aware of the master's potential loudness. As discussed in Chapter 3 'The monitor path', loudness can be observed on an RMS meter or more recently Loudness Units (LU) meter. I personally assess RMS as a more accurate visual impression of musical loudness. Though LU is clearly more helpful for mixed media and the balance of spoken word and music because of the weighting of the measurement of RMS. It is bias by the bass roll off and lift in the treble in the calculation of power, and this makes it less helpful to achieve an observation of the musical power in my opinion.

That said, the majority of normalisation systems use LUFS standards for mixed media systems. For example, a user on Apple Music might stream music, play audio sourced from their own uploaded CD collection, and listen to a radio show or popcast. A variety of levels of outcome which LU will bring into a closer range if Apple's 'Sound Check' is turned on. Having a meter to observe this outcome is a requirement in the creation of a modern master as much as a reading from a standard RMS is required. Equally being able to view True Peak (TP) and standard sample peak measurement as first set out for use in CD Audio is a required context.

Being able to view all these measurements simultaneously helps to gain a clear understanding of the master's loudness and peak level relative to clip. But remember peak observation and average level are not related other than the difference between them is the dynamic range overall. The difference between the maximum measured peak level and the top of the system (0dBFS) would be the available headroom.

CD audio peak standard is a straightforward thought process about how to define a clip. An audio sample in 16 bit will have a dynamic range of 96dB, of which every gain step of the 65,536 possible values per sample could be utilised. This includes all the available dynamic range including and between 0dBFS and -96dBFS. But the signal cannot go above 0dBFS, but even if the audio is at 0dBFS, it is not a clip. This dilemma was solved by the principle that a set of consecutive samples at 0dBFS would constitute a clip being activated on a system. For CD audio this is defined by three consecutive full scale samples equates to a clip. On some meters this definition can be changed to suit our needs. For example, the RME Digicheck metering can be set to a value from 1 to 20. If set to 1 or 2, a more reactive observation of potential clip is being observed. Remember, 0dBFS as defined in digital audio does not mean all DAC's are as accurate as each other at translating this to analogue. Or all DAC's would be the same.

True Peak is a more accurate way to observe the potential reconstructed peak on a DAC and the potential peak error when converting lossless audio to lossy formats. To achieve this, the audio is upsampled to improve the resolution of the view of consecutive full-scale samples. Doubling the sample rate from 48kHz to 96kHz doubles the number of samples, meaning if consecutive full-scale samples were set to 1, it would actually be a resolution of 0.5, double again 0.25 and so on. It is not unusual for meters to upsample five times to achieve accuracy. Measuring these 'inter-sample' peaks enables a meter to reconstruct the potential peak outcome up to +6dBFS above 0dBFS. The point of this observation is to make sure this does not happen, the audio should not evidence clipping at TP. Even though this is not really an actual clip in the DAW at best resolution in playback, it could be at some other point in the digital audio playback when consumed. All this means that adhering to TP at the end of our transfer path a TP limiter is required to upsample the audio and control these potential inter-sample peaks in the normal audio samples of the system resolution set.

There is no harm in observing TP outcomes as long as what is being viewed is fully comprehended and the engineer is not just trying to match a target level. The best way to work out what is effective is to observe some appropriate audio material – your references. As known, these sound superlative and must have been controlled effectively in their dynamic range and peak.

Practical setup of a transfer path

As the name transfer path suggests, the idea is to transfer signal from one point to another. In the earliest days of formats, it would have been tape to lacquer cutting, a purely analogue process which meant the audio quality would have been monitored during the transfer as it was in real time. Moving through the decades, tape was also transferred to digital tape for CD manufacture. Again, a real-time operation of listening, this progressed to digital file transfer to CD. Developing to digital only, the same thoughts about transfer in real time still apply, listening to the printed outcome, not just hitting render. If rendered off-line, the master still needs to be listened to all the way through. This just takes more time than listening while the audio is printed.

There are plenty of ways to set up to achieve transfer in this modern context, whether digital source to analogue path return to digital capture on a separate DAW, or even the same DAW. Digital source through digital interfacing loops back to the same DAW or transfer to a capture machine. Or simply use internal DAW bus routing to record the internal path outcome dependent on your DAW's capabilities. These potential paths can be observed in figure 10.1.

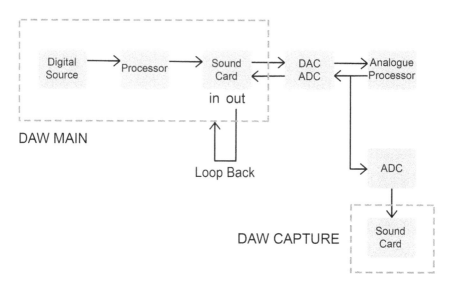

Figure 10.1 Transfer signal paths

All of these options need careful consideration of the bit depth potential, conversion and choices about sample rate depending on the source. When transferring between one aspect and another, normally something is happening sonically which requires our knowledge of the system's limitations. Whatever our choices, being able to listen to the audio while printing is helpful in many regards, something I feel should be done in any field of audio. Listening to it, not assuming the computer is doing its job. At some point, someone else is going to listen to whatever you have rendered. If you have not heard the potential error/s made, by then, it is too late. I cannot overstate how critical it is to listen to what you have created in full.

Practical monitoring path setup in a DAW

After establishing an effective transmission of the audio through the transfer path and a method to capture the print, it is also necessary to consider practically how to listen to this effectively.

To analyse and study audio regularly, there is a need to control our work environment to assist in comparative analysis. The most important contrast in mastering is listening to the original source without bias. Bypassing effects in the path is not an effective methodology in a modern context because a system often suffers from latency in this process. This switch is not smooth during the A/B comparison. A separate track that can be auditioned back without any processing is the most effective method. If this is sample-aligned with the source track, it can be simple to A/B between the original without transfer path processing and the source using X-OR type solo mode or toggling solo on a muted track in solo exclusive mode. Equally if you have another monitor input on your monitor controller, you can A/B at this analogue point in the monitor path, assuming you have two matched DACs. The advantage of this is that the DAW is buffered during playback, meaning the audition in A/B is seamless and everything is just playing as normal. Personally, I prefer this method because it makes very similar observations easier and seamless.

Having the two tracks aligned in the DAW also facilitates the creation of a sum/difference to hear the effect of the processing difference or to test an audio path in the digital realm. A polarity switch in the original track facilitates this. As A/B'ing, listening to either/or, is the normal comparison, the polarity can stay engaged all the time to allow faster workflow. Alongside this, our reference material can be placed on another track with relative levelling to achieve equal loudness comparative at any given point. It is then possible to listen to the original mix compared to the processing applied, and the reference material all levelled for equal loudness.

There is no correct way to set up, but to master, you should require each of these outcomes during the process. The most critical condition to achieve is equal loudness in any comparison at a point in the transfer path. In levelling back to the original mix, it means there is always ample headroom to be working with during the mastering process. This changes your focus towards the sonics to not worry about clipping or some ill-informed notion of correct headroom in digital.

Summary

There are four important considerations in construction and maintenance of the transfer path. First, be able to meter and observe all the I/O points in the chain to maintain loudness level and equal loudness. Maintaining best quality in the signal path at the reference level avoids any potential clipping. Secondly to metering, meter in different standards simultaneously to give a full context to loudness and peak and dynamic range overall.

LUFS, Peak, TP and RMS are all important to observe on the final master output. Third, adhere to digital clocking protocols to negate any jitter and achieve the most effective resolution for the system. Fourth, use upsampling to your advantage to achieve the most accurate processing during the transfer print.

Practical approach to dynamic controls

Knowing your tool types

No matter how much theory is observed and principles directed towards which tool type would be ideal to use in a given scenario, in the end, engineers only have the tools available to deliver an outcome they are ideally trying to achieve. Practically, this means any processor to be used has to be fully interrogated to understand any spectral artefacts it may create. As discussed in the formative chapters, tool testing and sum/difference are key to gaining this knowledge. If the primary, secondary and tool type outcomes a processor creates are fully understood, this knowledge can ultimately be used to your advantage. As an example, this should include how the dynamics tool knee works. Every dynamic processor will have one, even if this is fixed. Each dynamic tool has a different transition to the attack and release, and this could be exponential in shape, linear or logarithmic and so on, the same as would be observed in adjusting a fade curve. Comprehending how your tools function in all aspects will make sense of the differences between them. For example, in the time settings, one compressor could have 2000 ms release, and another sounding similar in response in length could be half of that in ms because of the envelope.

Nothing comes for free. If you want to become effective at using dynamics you need to start investing time understanding how the tools at your disposal work. The manual will often be a helpful source of knowledge, after all, the people designing its intent will know better than anyone how it potentially could be used. In starting this investigative process, it will also make you aware of what you want out of a dynamics tool, and not what someone else suggested. Confidence in comprehending how tools function and sound also help you to know what sonic properties your next purchase should have to compliment your current pathing possibilities.

Secondary outcome consideration

When adjusting a waveform's envelope shape, the tone is also changed, the two aspects are intrinsically linked. When creating more sustain (upward compression), there is an increase in amplitude perception of bass at equal loudness. When reducing the impact of a waveform transient, the sound will be dulled (downward compression). The more the focus is shifted to the actual peak of the transient (limiting), the greater this apparent dulling will be. When lifting the transient (upward expansion), the audio will become perceptively brighter because of the increase in the attack of the envelope. These observed changes in tone actuated because the primary outcome of dynamic control must form part of our appreciation of how to use these secondary outcomes in tonal change to our advantage. Achieving these secondary outcomes from our dynamics processing means the need for EQ is reduced. It can be totally negated during some mastering if the effect of tool type is used well. This assumes the mix is positive in the first place where corrective processing is probably not assessed as part of the path.

Considering tool types' effect on the tonal shift of the envelope, when engaged in downward compression using a valve process, it will add saturation and harmonic content brightening the sound. This can often negate the dulling from the downward compressions secondary outcome. This is why many mastering engineers enjoy the Manley Variable MU, because when applying a couple of dB of reduction, the overall tonal impression at equal loudness would be a little bit brighter. In contrast, a full band clean digital compressor would sound softer in tone in direct comparison at equal loudness. Clean means no tool type outcome is present and the secondary outcome would be dulling the tone, making it even duller with more bite applied in the attack. Obviously, this could be corrected by pushing in the compressor with a high shelf, a low 3dB/Oct achieving a smooth lift, more of a tilt to the treble starting at 2 to 6kHz to reverse the dulling. This means there are now two tools and two possible colourations, as opposed to one tool. It could also be possible to use our clean digital compressor and a soft clipper in a soft, not hard, mode to achieve harmonic distortion on the peak elements. This would create a similar effect to a valve compressor in principle. The TC Electronic Brickwall HD or Sonnox Oxford Inflator can both achieve this more subtle addition, and it is best to negate any volume change from the plugins to clearly hear the true effect of the saturation applied. Most saturators should be able to achieve this addition if used appropriately. With this addition, we would now have three tools and three potential colourations as opposed to the one valve compressor. This is why consideration of tool type and secondary outcome are key to a successful approach to mastering.

As another example, if the assessment of a track sounds too bright and too dynamic, it could be correctly concluded that the use of downward compression and EQ to reduce the tone would be effective. But if selecting a valve compressor and valve EQ, it would add harmonic distortion (brighten) from the tool type outcome, meaning more EQ would be needed to correct, adding a bigger change to the audio overall. Not the best outcome. The audio would have been improved in comparison to the original as we have responded to our analysis, but not as well as if a clean compressor and EQ had been used. Clean tool types applied to the path would need less EQ, as the secondary outcome of applying the compressor would be to dull the audio. With critical evaluation of the attack setting, it might be found this single compressor was enough on its own. Stacking clean compressors and reducing the attack on the second unit relative to the first can help shape the transient to give the required dulling with the correct amount of bite to the compression. This makes for a reduction in the overall processing and refining the path once more. What counts is what you leave out through correct tool selection and not the processes applied in creating the ideal.

Just because classic analogue tools have character, does not mean they should be used. Our tool selection and type must be driven by our analysis and its contextualisation by achieving an appropriate response in processing and taking into account the potential primary, secondary outcomes and tool type.

The link between threshold and ratio

There is a straightforward observation when initially setting the ratio for the best starting point with compression. The lower the potential threshold, the deeper it is into the signal, the more its gain reduction will be active because of the ratio. This means the ratio needs to reduce as the threshold goes lower into the signal, otherwise the audio will become over compressed. Conversely, the higher the threshold, the less the signal breaches and the larger the required ratio needed to achieve appropriate gain reduction. The same premise is true of upward expansion. If the threshold is breached by a regular strong signal, the ratio will push the level dramatically if the ratio is too large. If it only breaches by a very small amount, a larger ratio is required to achieve a considered reaction. For example, in de-essing a vocal in the mix, it is the loudest aspects of the sibilance in the frequency band that are being targeted, the ess's. This means the threshold will be high to avoid triggers on the hi-hats or ride, or the other parts of the vocal treble range. The threshold is only breached by a small margin, and a large ratio is required to achieve enough reduction. 5 or 6:1 would often be a good starting point for de-essing a mix as the threshold will be very high to avoid these mistriggers.

In another example, in compressing a whole mix to control the macro dynamic, the threshold would be deep into the audio to ride the average level, not triggering much in the quiet sections, but activating in the louder sections of the song. A high ratio or even a medium ratio of 3:1 would cause potential pumping, but a ratio of 1.25:1 or 1.5:1 will react to the mix effectively without negative fluctuations. In researching compression tools, you would note this is the fixed compression ratio on hardware mastering compressors like a Manley Variable MU set at 1.5:1. As a theme you would observe that mastering and good mix bus compressors equally have low ratios, if not even tiny ratios as low as 1:03:1, as opposed to mixing compressors with lowest ratios starting at 2:1 if not 3:1. Lower ratios enable the compressor threshold to be deep into the audio without pumping and smoothly compress the mix. Equally it means the most optimum trigger from the threshold can be achieved without having to compromise because the ratio achieves too much gain reduction at points. Remember, threshold is the trigger, and ratio is the control of the relative change in amplitude.

Practically in controlling a micro dynamic, it is often to adjust parts of a mix to improve the bounce and cohesion. This would make the starting ratio more akin to the mixing environment of 2 or 3:1. The threshold will not be as deep in the audio as our previous macro example. A higher threshold equals a larger ratio. The more analysis drives the transfer path design, the more straightforward it becomes to set the control close to the ideal for a given process because it is known how the processor is meant to react to the source. This saves critical listening time, keeping our listening perspective fresh. Being close to the ideal to start with makes our response a quick tweak to the threshold, attack, ratio and so on to achieve the optimum. This is in contrast to turning a processor on in the path and moving all the controls around to 'find' a sound while listening throughout this process. Not only does this take a lot longer but we are inherently losing our critical perspective.

The more consideration you can give to the processor initial setup before engaging in listening, the more accurate your analysis and evaluation of the outcome will be. This method in approach is even more crucial with dynamic processor control.

Adjust one, and you change them all

Dynamic control is one of the hardest aspects of audio engineering to achieve proficiency in. Anyone who says 'I know everything about compression' probably knows very little. I would consider myself very knowledgeable and practically skilled, but I have by no means engaged with every outcome. Musical form is vast, hence there are so many potential dynamics possibilities you could encounter. I learn something new and observe a differing nuance every time I engage in processing. You have to work at your art, and be humble to it. The main aspect to comprehend in utilising dynamics is that all the controls affect each other – adjust one, and you change them all.

If you reduce the knee towards hard, the threshold trigger scope will reduce, meaning the amount the threshold breaches in range is smaller. The ratio's effect will appear more variable and aggressive, and attack will appear to bite more and the release will seem quicker and so it goes on! Every control affects the others. Comprehending this interlink is critical to get to grips with how to change the audio in the way required to shape the amplitude. One way to start to develop this skill is never adjusting the dynamic controls variably when listening to the audio. It is important to consider the initial setup of the attack, release, knee, ratio and crest/SC. Set these parameters to where it would be anticipated the requirement in response would be from our analysis, i.e., how the compressor is intended to change the sound. When the music's playing, and dropping the threshold to achieve a trigger, the observation in gain reduction and sound adjustment feedback from the compressor will be somewhere along the lines of what was expected. A quick evaluation in how it has affected the signal can be made and observed, thus what adjustment would be needed to achieve our goals. Too much squash, but a good trigger, you need to reduce the ratio; if still too much bite, then lessen the attack or widen the response of the threshold by increasing the knee, which reduces the variance of the gain reduction and so on. In some ways this is why some engineers sing the praises of minimalistic dynamics tools with only simple control such as input/makeup/reduction. They do not have to make these interrelated decisions because the adjustments are auto or fixed in the tools response. This is useful in one way, because quick decisions can be made, but very restrictive in achieving the kind of outcome required in complex programme material during mastering. To achieve positive dynamics control, it is critical to embrace the functionality of the available tools. I would say one of the fullest featured compressors in this regard is the Weiss DS1-MK3. If you want to really start to comprehend the purpose of each control used in a dynamics processor, I would look to engage with this tool. Even if it is just reading the manual. Designers and engineers do not include features on a tool for the sake of it, it is because in the real world of daily use they are genuinely helpful.

Release and loudness

Backing off the release equals less compression over time. Less compression means more perceived loudness as the amplitude increases relative, and as known, louder sounds better. I have seen engineers backing off the release, listening, and back off the release some more because it is sounding better, until there is hardly any release, but plenty of gain reduction on the meter (looking and not listening). This very fast release has really made an envelope follower, a mini distortion box as the compressor is going in and out so fast. The gain reduction is just tracking the waveform around as if it is the carrier signal for a modulated distortion. This also adds harmonics potential to the compression outcome, not only does it sound louder by amplitude, but also brighter. This effect is a very fast 'flutter' or 'chatter', a term more commonly associated with gating. But a gate is not opening or closing which would be obvious. In this case, the flutter effect is often masked by the rest of the main signal.

Compression is difficult because it is critical to be aware at all times of the effects of loudness when manipulating the amplitude. Think about what is trying to be achieved, and not moving variable controls all over the place will help develop skill. Stepped controls, or working in steps helps to consider what is actually happening to the signal and our perception by making clear A/B's. Comparing 10 ms attack and stepping to 8 ms and back again is a clear comparison, and the change in the dulling of the audio can be heard when the attack is faster. Using variable controls, a move between 8 and 10 ms cannot even be accurately achieved. What you are listening to is the transition change in attack over time when you sweep from about 8 to about 10 ms. This is not helpful in critical analysis especially when listening to complex material rather than just a single instrument when mixing.

The attack control is more finite. Most engineers will clearly hear the difference in 0.05 ms in the bite of an A/B with a mix which has a complex envelope. It is not unusual to want a very fast attack to grab the transients and control the treble below 1 ms. Making an A/B between 0.3 and 0.25ms or 0.8 and 1ms are not the easiest values to set accurately with a variable control, let alone go between them. During mastering accurate control is required for our ability to critically evaluate and make the best selection for the musicality of the sonic.

Observing the gain reduction

I cannot overstate how important the observation of the gain reduction meter is in truly starting to effectively control dynamics. Viewing the gain reduction meter and seeing its full recovery during the music's transition is critical. Any signal being adjusted in dynamic range must have a lowest sound and a highest sound in both peak and RMS. To achieve downward compression, those quietest aspects as a minimum should not be affected by the compressor. Because if they are, the compressor is just turning the amplitude down overall and making the sound more coloured by the process applied. It is critical to observe full recovery of the gain reduction meter when the audio is playing. This does not mean where the music has a pause and the gain reduction releases back to zero. This recovery needs to be seen when the audio signal is present. This means where the audio signal is at its highest point, there is the most gain reduction and no gain reduction where there is still a signal but in its lower aspects. Thus the dynamic range has been compressed, applying contrast to the original signal and not just turning everything down. The more this is observed in practice, the more targeted the processing will become and the more effective and transparent the outcome, hence being able to colour the audio with better precision. There are no negatives to this, it will only make your dynamics processing better.

This contrast is even more important in upward expansion and is absolutely critical to achieve effective parallel compression (upward). All these techniques are discussed in detail in Chapter 12 'Practical approach to dynamic processing'.

The lower mids

The nature of reducing the dynamic range inherently means an increase in the perception of the trough in the waveform, whether this is from an upward or downward process, or both. This increase in sustain changes our perception of its loudness, but it also adds to the density of the frequency balance. This generally manifests in the low mid, anywhere from 150Hz to 500Hz. As the audio gets denser, the more this area will need addressing potentially post the dynamics processes. The action of backing out a little of the low mids can in principle help to open out the master, separating the bass bounce from the rest of the music's chordal and melodic elements. This would probably only be about half a dB, not a large scoop. As a general theme, it is best to avoid negative EQ but using a linear phase EQ will minimise the smear on the audio frequency balance. Being conscious the build-up will happen means you can be listening out for it as the path develops and mitigate some of this outcome by focusing each process applied. An attempt to address the imbalance could be made by driving in to the dynamics pre, with positive EQ lifts to the sub and highs, though this affects the following thresholds and frequency balances in the path. A little post EQ will still probably be required if there is a lot of gain reduction. The more the reduction in dynamic range, the more this outcome will become apparent. It is often more effective to just tweak the frequency balance post main compression before skimming off the peaks in the last stages of the processing path. It all depends on how much dynamic range reduction is required overall. Making the potential cut pre would seem logical, but will undo the positive processing achieved and is likely to thin the music. Cuts at this point in the transfer path are best avoided if possible. It is better to apply a lift in the bass and treble to soften the effect of the lower mids. This idea is discussed more in Chapter 13 'Practical approach to equalisation'.

Ganged/linked controls/linked gain reduction

The ability to link or unlink controls should not be confused with linking and unlinking the sidechain on the dynamics tool. As discussed in the previous chapters the need for separate channels in mastering is driven by observing the stereo mix in its composite parts Left and Right, as well as matrixed in Mono/Mid and Difference/Side. By linking those observations in analysis to a discrete path of processing in LRMS/Stereo, more effective outcomes will be achieved. Being able to have independent controls means they can be set differently for each path. Considering a compressor receiving the same information in each discrete sidechain/key input, if the controls are unganged (not linked together), there are now two sets of control to potentially trigger both sides of the audio signal so there are now two compressors active in one unit. This is because most compressors when 'linking the sidechain' are actually linking the outcomes of gain reduction. The controls set on one side give an outcome in gain reduction over time. This is sent to the same gain reduction trigger on the other channel of the compressor. Setting the controls on the other side differently will send its outcome to the other channel gain reduction circuit, thereby achieving two differing layers to dynamic control from one process. But careful consideration and testing are needed to check how the tool in use responds. A Neve 33069 and Weiss DS1-MK3 will work in this way, as will a Manley Variable MU, unless you are changing the fixed ratio switch on one side. This does not translate its dB reduction to the other side. Both need to be set the same to achieve the effect in balance.

In practice this can be a useful feature that I have used during analogue mixing when bus compressing a group of elements with fluctuating outcomes. Having a longer release at a high threshold and the normal response lower helps contain the louder elements. The ratio is lower on the first trigger level, and higher on the second. This works well when there is only one Neve 33069 in the rack to use. But I have not used this during mastering, mainly because we are in a digital age and the limitations on gear experienced in the days of analogue and digital hardware are no longer an issue. If I need another response in dynamics, I can just add another plugin. Now it is possible to insert two, three or four in the path to gain more control and accuracy with each process. You must always be aware of the tool type in stacking any tools to avoid more and more colouration.

Sidechain compression

Many dynamics tools offer the ability to access the sidechain (SC) of a unit to manipulate this audio bus. Often in mixing, this is used to trigger one sound with another for rhythmic exchanges. Simply pocketing the dynamic range and frequency to unmask achieves a more even sound balance, to be denser and more compressed. It is not unusual to want to trigger the SC during mastering with one aspect of a stereo mix from another even in this complex programme material. Though initially the basic principle that an SC is just an audio bus needs to be understood, meaning it can be manipulated like any other audio channel, for instance, a reverb could be inserted if inclined. But thinking about the simplest tools to manipulate a bus, our focus should be on how filters, polarity and delay could be used to our advantage. This principle of the signal flow in a dynamics tool can be observed in figure 11.1.

We will look at filtering first. A dynamic tool works by audio breaching the threshold, the louder the amplitude the more it will make this transition. The wavelength of the audio changes the amplitude response, the lower the frequency the bigger the potential amplitude and eventual trigger of the threshold. By changing this relationship on the SC by applying a HPF, the lower frequency will have less impact on the response of the trigger threshold. A HPF at 100Hz 12dB/Octave would mean a full band downward compressor would trigger more from the other main elements of the

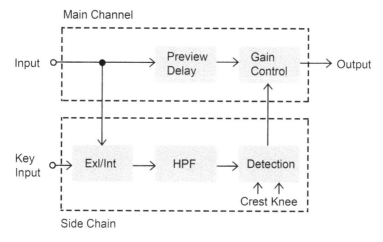

Figure 11.1 Dynamic tools sidechain and main channel bus signal path

mix, vocal, snare, guitars/synths and not the kick and bass. The fluctuation in peak to RMS will also increase as the sustained elements are reduced.

This application is not the same as split band compression where only the audio passed through the filter will be processed. Set the audio processed to full band with the HPF only on the SC so this sound is distinctly different, keep the original audio mix balance intact at any given moment in time. 100Hz to 150Hz with a 12dB/Oct filter is generally interesting. I personally find this helpful where I want to push into the vocal or snare, but not target the vocals to reduce them relative to everything else by splitting the audio in frequency bands which would make the mix disjointed. Once the bass is removed from the SC, the compression threshold will ride those louder aspects of the vocal and snare in a positive manner, especially if the crest is set to RMS. You will get a musical outcome with minimal invasion into the original mix balance, particularly with the tone.

Coming back to the original thought of filter and delay, delaying the SC is just the same as backing off the attack in principle. But as the main audio process path is a separate bus to the SC, in delaying that, a look ahead is created (see figure 11.1). If you have a dynamics tool that does not have a look ahead but does have an external SC/key input, a simple way to achieve look ahead is to have the dynamics on the main audio channel and have a separate copy of the audio sent to the SC/key input. Moving the position of the two audio tracks changes the perspective of the difference. If the SC is ahead of the main channel, a simple delay/look ahead is achieved, while still maintaining original sync to any other related aspects. If our attack is a step control, this can be a useful way to achieve a more finite response from this control, or effectively make the knee seem harder as it is potentially more reactive to the signal over time.

Split band principles in application

There are two different thoughts here. Either target a specific frequency area for an instrument in the mix or look at a broad aspect like the whole bass, middle or treble. Our focus is either corrective or looking to affect the tone of a large frequency range in overview. The point in overview is to change the relative frequency balance, but apply some dynamic control in the process. When engaged in this, it is best to be clear about the filter shape. The roll off of the filter needs to be soft to achieve an effective lift or cut in the EQ balance without it sounding like the audio has been hacked in two as it would be with a tighter/steeper crossover. If it is soft, the shape for the high and low bands will effectively be a shelf. The TC Electronic MD3 has these slow roll off curves as an example, whereas a UAD Precision Multiband has a steep/aggressive roll off. This broad targeting of the bass or treble can be a really effective way to tilt the audio spectrum in the same way as a high or low frequency shelf on an EQ. But if achieving this on the compressor dynamically, there is no need for another processor in the path because the makeup can be used as an effective gain control. Less processing in the path is often more of an effective musical outcome for the audio.

On the other hand, when trying to target a specific aspect, apart from using a multiband with step roll off on the filters such as a UAD Precision Multiband, it is also important to make sure the focus of those cut-offs are in the right point in the spectrum. When listening to a solo of the band you are processing, it is almost impossible to achieve the correct crossover because the sound being processed is being observed. A better approach is to solo the band next to the aspect to be changed. When listening to the band next to the one to be processed while moving the crossover point, it is possible to hear the aspect to be targeted because it starts to appear in the audio band, hence setting the band to perfectly capture the sound required. The same applies to the other side of the banding as well if it is a bell shape. If there is only a single high or low shelf crossover, there will only be a single crossover to set.

Often when engaged in this evaluation, the aspect to really listen for is not where the sound to be processed appears, but when it can be heard dynamically. After all, the rationale for applying a corrective multiband is because the sound is overly dynamic in the context of the mix, for example, where the observation is an overly sibilant vocal. The sibilance will appear across a broad range of the higher frequencies. The target is not all of the sibilance, but the aspect of it that is overly dynamic. In listening to the adjoining band to that of the sibilance, listen for when the sibilance becomes dynamic to set the filter and not just hearing sibilance. This is normally quite obvious when focusing on that dynamic aspect. It is not unusual to have more than one frequency area to target in this way with sibilance. The Weiss De-ess is a very effective tool to address in one processor as it has two variable target frequency bands but you need to listen to aspects on either side of the sibilance to find the band point because it is focused on the processing outcome only. The UAD precision multiband can listen to bands around that to be processed. Another effective tool is the TC Electronic System 6000 De-ess algorithm, though you need to use the side chain monitor to find the range as it is not possible to hear the other sides of the processing band. As with the Weiss De-ess, you will be accustomed to this principle with practice.

Mid Side application principles

MS compression is mainly about the SC being linked or unlinked. In setting a Mid Side matrix around a stereo compressor with the controls ganged (linked together), the through put signal will now be in Mid Side. The nature of most music means the main power of the signal in both loudness and peak would be focused in channel 1, the Mid. The trigger for any gain reduction would come from channel 1 and be shared with the linked SC, and the Mid would be triggering the Side most if not all the time. This can be useful when imposing the timing of the main rhythm on the Side elements. This might be guitars or synths sitting in a hard pan and the majority of their signal is in the Side. Having the SC linked means the trigger affects all the signal path. With the crest control set to peak, a trigger from the main percussive elements will be achieved, generally with an effective trigger off the drums. To enhance this focus further, ungang the controls and as long as the trigger is coming from the Mid, the controls of channel 2 can be adjusted independently. Depending on how the gain reduction link works with the tool, as discussed above, it should be possible to increase the ratio to work more aggressively punching holes in the sustained guitars or synth to give greater bounce making more room for the drums in the bigger picture of the song.

Setting the controls to unganged and the SC unlinked from the outset, the Mid and the Side can be compressed or expanded independently. This is something used in cutting a lacquer to control the energy in the lateral motion (Mid) and the vertical motion (Side) to avoid break-through during a cut. This is why some historical compressors are labelled Lat/Vert or what is now commonly referred to as Mid/Side.

In a modern context of delivery I find this mode of MS compression most useful where the mix is focused on the mono. Compressing this element helps to change the balance in the stereo image as the Mid outcome has less amplitude than the Side. Assume the makeup gain on the Mid is at unity and the main output gain is used to rectify the output signal back to equal loudness. More Side overall and less Mid relative can be enhanced on the same tool by expanding the Side element to increase its punch with the attack to the fastest and/or by adding look ahead to get in front of the transient and release at 50–100 ms to create more width definition. Compression on the Mid and expansion on the Sides gives a wider image by default of the gain structure change. But this is driven by the music's dynamic and not a fixed amplitude boost or cut. More musical rhythm is created from the mix.

In some instances when correcting a mix, there may be excessive cymbals overall, but it is hard to achieve a trigger with just the frequency selected on the multiband stereo. The Mid element is likely to be more complex in content, where the Side is probably a more effective trigger. Switch the compressor into Mid Side and set the threshold of trigger from the Side element. With SC linked, a clear trigger can be achieved affecting both Side and Mid. The threshold on the Mid needs to remain very high to avoid mistriggers from the Mid.

There are a myriad of possibilities when starting to use independent channels, SC and an MS matrix. The best and most appropriate method should be driven from analysis, not just the premise the facility exists. Everything should always link back to analysis.

Hearing the envelope

In developing your ability with dynamics processing, the use of sum/difference with a clean tool to hear the gain reduction being removed or added is a vital part of developing correct use of the envelope with dynamic tools. Being able to hear the actual signal being changed in isolation will help to set the attack and release precisely, to match the envelope shape that is needed. This is especially true for the release, this aspect is more challenging than the attack where the bite or dulling of the transient is more obvious in open playback. Listening to the sum/difference makes the release clearly audible where it recovers. Too long and the next musical transient will be heard, too short and the gain reduction recovers before the length of the target is complete. I have found that it is an especially useful observation in the application of de-essing, when using a linear phase multiband with the frequency range focused on the sibilance. The recovery and release time required can be clearly heard to achieve a timely reduction on the 'S'. This is often much longer than you would imagine in release. A more musical and often smoother effect is achieved. Some tools like the Weiss De-ess have this functionality built in to monitor what is being removed.

Utilising a sum/difference polarity switch against the original is also a positive way to 'hear' the effect of a limiter or soft clipping function. Adjusting the level of the amplitude to achieve as close to null will also refine your ability to observe the difference because many of these tools add some gain difference. When switching back to the full processed signal, it is then an equal loudness comparison to hear the actual change in the overall aesthetic of the music.

Practical approach to dynamic processing

Dynamic principals

In starting to control the dynamic range of complex programme material, it is not possible to achieve the outcomes expected from a modern master with just one type of dynamic reduction. There needs to be appropriate morphing of the presented mix with a set of stages to achieve the control that would be considered mastered. As a rule of thumb, achieving more than 2dB of gain reduction on average out of an individual process in a mastering context is probably going to be obvious. Which is fine if that is the aesthetic choice, but often if the mix already sounds good (as it should), there needs to be a more considered reduction to avoid the dynamics negatively affecting the mix's positive sound balance. A mix could have a dynamic range of 20dB, a master between 12dB to 5dB depending on genre and measure used. From that perspective, it would mean four to seven different actions to control the amplitude would be required in the transfer path to achieve a shift in dynamic range without negatively affecting the good sound balance present. In applying a little bit at a time, it is the summative effect of this processing that will give the dynamic range restriction anticipated without any negative artefacts.

This does not mean just stacking compressors and limiters until the reduction imagined is achieved. Each change needs to target a different part of the envelope. As one aspect is adjusted, it opens up another to the possibility of processing. For example, a limiter could be applied at the start of the chain to skim off only the maximum peaks, not to get every loud transient but the loudest ones. This sets a headroom for all post dynamics processes creating a known range in dB between the following processor's threshold and

DOI: 10.4324/9781003329251-13

system clip at 0dBFS. Equally the dynamic range between the audio peak and RMS can be observed and the knee is easier to predict in the potential range of response of the ratio control. The limiter would only affect the audio for a very small portion of the actual music over time with a maximum change of a couple of dB. This does not mean the intent is to increase the loudness overall but rather to maintain correct listening/reference level. The outcome is still the same perceived loudness but with a controlled maximum peak level.

Applying another limiter now would target the same aspect and would sound obvious because it is basically the same as putting more reduction on the first process. When targeting the micro dynamic with a downward process coming behind the main attack of the transient, it will be reshaped. Thus, another limiter following would be shaping a different impression of the transient as much as an EQ to boost the treble (if required from analysis) could change the sharpness of the transient. Alternatively, a soft clipper would make this change more explicit on the transient before another process because it is distorting and shaping the transient below and above the threshold set. What is critical is to make a change to the envelope shape before applying the same type of process once more. There always needs to be a different target in the envelope shaping with each following process.

Coming back to our mix dynamic range to achieve a mastered outcome, I might expect to apply a limiter, upward compression, downward compression, soft clipper and TP limiter in terms of just the dynamics. Given our 2dB guide on each, that is a potential of 10dB. Through adding upward expansion to raise the transients in the Side element and a little manual compression in the quieter parts would mean a potential 14dB difference could be achieved. Probably a few dB too much, because some of these processes would have a lighter touch or would just not be needed. The key is to address aspects in stages, and not try to achieve it all with a couple of processors. That approach will not work unless the audio is already very compressed from the outset such as an EDM track. This will inherently be compressed as the original dynamic range is dedicated by the velocity setting in the sample playback. In comparison to a fully acoustic recording, it would already sound almost limited and densely compressed. Knowing your dynamic range and the potential target dynamic range will be helpful in starting to manage this appropriately.

As a theme I am not trying to 'hear' my processing directly. I am looking to address the mix dynamic as transparently as possible to achieve the desired dynamic range with the audio sounding unmasked, balanced in tone and not processed sounding in any way. I want to achieve the good 'glue' effective dynamic control creates without hearing its processing in the music.

Downward compression

One of the biggest differences between mixing and mastering compression is in the application of ratio. In mastering, a low starting ratio of 1.5:1 to 2:1 is usual, it is less rather than more. When looking at an individual instrument in a mix, the contrast in dynamic range is generally larger, meaning there needs to be a greater positive ratio 2:1/3:1 or more to achieve enough gain reduction to smooth the audio. Remember that the point of compression is to reduce the dynamic range, and it is important to be conscious of how the gain reduction is responding. In reducing the dynamic range, the louder aspects should be triggering the compressor, but the quietest aspects should not. There is contrast in the gain reduction, making the dynamic range reduced across the excess dynamic.

It is critical to observe the gain reduction meter effectively when compressing anything at any stage of the production process. If the gain reduction does not recover (turn off) during the audio playback, and it is on all the time, the audio is simply being turned down. The point of compressing something is that it has already been analysed as overly dynamic, which means it must have a loud part and a quiet part, otherwise an incorrect assessment has been made during analysis. If there is a loud part, that is the aspect in focus to be reduced with the compressor leaving the quiet part unaffected. This is in part why dynamics are more transparent than EQ. A dynamics tool is only on for some of the time (momentary) and an EQ is on all the time (static).

When starting to reflect on approaches to full band downward compression, the directives can be split into two aspects depending on what is trying to be achieved with the envelope shaping. This focus will be around either the macro or micro dynamic, which should be driven from observations during analysis, and not without clear purpose just applying a compressor and seeing what happens.

Approach to a micro dynamic manipulation is normally a musical response to the audio sitting within the rhythmic beat and where the transient sits in relation to this. If focusing on controlling the brightness/attack of the track and holding in the transient, use a pre-process to hold in the width of the transient at 25–35 ms and a release to sit in the gap before the next transient set at maybe 100–250 ms with a medium paced song. This pre-shape can be followed by a compressor with much more bite at 5–10 ms or even sub 1 ms if the audio is not overly dynamic. A similar release setting to the pre-process to fall in with envelope shaping is already created. Both these compressors would have a soft ratio as a starting point at 1.5:1 or up to 2:1. These processes both need careful A/B back to the original to not over dull the audio. Our evaluation in analysis should be the guide to the amount required. Look to set the second compressors attack so the transient still breathes. Too fast on this second compressor and you will kill the drive of the music. If the music was very transient, the attack could go less than 1ms, even as low as 200us. The most important aspect is to A/B, listening for where the bite is appropriate. The knee would be tight, at maybe 3–6dB or harder to reduce the potential range of the trigger. The threshold is likely to be higher than the first pre-compressor as the attack bites harder. This stacking can also be helpful pre-processing for a soft clipper to achieve a smoother dynamic response, especially when limiting the second compressor output lightly to set a maximum level to the transients from these outcomes that will now be shaped. In one way, this type of compression process can be thought of as a form of expansion because the distance between the transient and trough can often be larger with the compression processes only, punching a hole behind the transient to create bounce and impact while controlling the width of the transient attack. When the transient is controlled and uniform, it is possible to skim off the sharpest aspect without distortion and/dulling artefacts that would have been heard if just applying the limiter to achieve the same gain reduction overall. With a dense mix, the first compressor would probably not be needed because the music's transient shapes are already pre-shaped. Though if it is an overly compressed mix, you can push the compressor harder with the ratio to expand the beat by creating space behind the rhythmic transient elements.

This stacked process is likely to be more used with music that has a wide dynamic range in the mix, where the source instruments are real, such as country, folk or rock. With a more compressed musical source with pro-grammed material such as an EDM track, there would only be a need for the second compressor, if any at all, because the track is already compressed and the focus would be on trying to shape the musical bounce/attack and be less focused on controlling the initial transient overall.

The tighter the bite, the more the dulling. This could also be countered with a treble boost before the compressor, but the mix will sound more processed. Personally, I find this generally does not sound natural to my ear, though with some programmed tracks where more bounce from the rhythm is required, it can be effective in delivering clarity with the change in transient shape, plus the sounds are not necessarily 'real' anyway making the change in timbre less obvious than on a performed folk song.

When controlling the loudness in the macro, a much slower compressor response is needed to help ride the level. A Manley Variable MU is a good example of a slow attack and musical release in a macro mastering compression context. The fixed compression ratio is small 1.5:1, making the compression less reactive, and a very soft knee helps it ride the macro dynamic smoothly without pumping unless pushing the gain reduction too hard. An attack setting of F/G of approximately 40/50ms will allow the main hits of the rhythm to come through and the recovery (release) control set to the speed of the musical recovery. The longer the hold, the slower the response to the change in dynamic over time. Med Fast/Fast 400ms/200ms normally sits well with most pop music; if it is a very slow tempo, a longer Med (600ms) recovery will help. Setting the threshold on the loudest sections to achieve a couple of dB gain reduction maximum on average will mean the volume is being reduced in these loud sections. Now auditioning the quieter sections of the song, the gain reduction should be off or just minimal in response. Contrast has been created in the compressions gain reduction as the dynamic range in the macro has been reduced. The louder sections are quieter, the softer sections are the same. If there is no contrast in the song, it was already compressed in the macro, and this should have been observed in analysis already. This does not mean the micro dynamic is okay. If our analysis is correct it should point to how and what the outcome is for any other dynamic control.

This style of macro compression can be replicated without the valve 'tool type' on a clean dynamics processor like the Weiss DS1-MK3. Set the release slow to the same with average of about to 200/500 ms, again dependent on the pace, and an attack of 50ms to let the transient through. Set knee range in the menu to 30dB and 30/50% (9–15dB range) on the control. You could find that setting the RMS window to about the same as the release time will help for a smoother response by ignoring the peak trigger point if the compression feels too hard. Again the ratio would be small (1.5:1) as the threshold is low in the audio during those louder musical sections. We are trying to ride the louder sections down and leave the quieter sections with minimal compression. If you are still achieving too much compression on the quiet sections, reduce the knee range and raise the threshold a little but still maintain trigger on the loud sections.

When applying downward compression, one aspect to watch out for is the release time. If too fast for the music the compression is only really achieving an action of following the envelope of the original signal. But the gain reduction will be quite active on the dynamic tools meter. Though really, all we are doing is rapidly releasing from after the attack, and as the threshold is low it will go straight back in again. In an extreme case, buzzing the waveform would cause flutter as discussed in the previous chapter.

There are so many possibilities as to how to apply downward compression, the key is in starting from analysis and fully comprehending why it is being applied. What is it targeting and why and what is to be changed about the waveform. Responding directly to this envelope shaping will achieve positive outcomes as long as equal loudness comparisons are being made back to the original mix to hear how the envelope and importantly the tone have been changed by our processing.

Expansion

One of the more useful tonal tools in the dynamic arsenal is being able to increase the intensity of a sound over time. It is not unusual to be faced with mixes that lack punch and bite in the attack as they are too levelled or just lacking bounce. It can also be the case that in processing a considered amount of compression needs to be applied and now the music feels a little flat in its dynamic bounce, probably because the mix was too linear in its sound balance in the first place and lacked density. Expansion can fix these issues, though this would normally be in a frequency dependent aspect, like expanding the kick weight or the snare mid top to achieve more punch in the master. Equally, give a heightened attack to the top end in general with a broad dynamic HS. To achieve this, a dynamic EQ or split band compressor will be an effective tool to select. The Weiss, Sonnox or UAD precision dynamics are very positive in achieving this kind of bounce, for a really clean and precise sound with no distortion. For example, processing the main kick weight with a bell filter makes that aspect bounce to give more power to the kick. This does not mean tracking the kick with a really fast attack to turn the whole kick up, as this would only increase the overall dynamic range. The outcome needs to sound like the kick is more forward in the mix to have more impact and force to be louder, but not causing an increase in dynamic range on the already compressed sounding parts. This can be achieved by targeting the filter around the power of the kick. Rather than trying to push the attack, it is the length that needs to be extended in the sustain of the kick to give it a fuller density in the mix. In doing this, the RMS is increased and the dynamic range overall is reduced; even though upward expansion is being utilised, the outcome would be more compressed overall. Setting the attack after the main kick attack to 20–30 ms and the release quick behind this at 30–80 ms, with a low expansion level of 1:1.15 or 1:1.25, should give a pullup to the back end of the kick increasing its power and moving it forwards in the mix sound balance.

As another example, if a mix lacks attack in the Side element, using an expander can give more bounce pushing it forward in the depth of field and hence respectively widening the stereo image while increasing its transparency in the Sides without just creating a static amplitude change. This keeps the Sides more dynamic than a global amplitude rise, but with a higher impact in the mix balance to be more rhythmic. To accentuate this change, use a split band compressor, processing from above the low mids, at 400Hz up leaving any weight in the Side to stay the same, but it can work well on the whole Side element if there is not too much density and sustain in the lower mid element of the Side. Either way, a fast attack and even with a look ahead to capture all of the transient can be used. A release to suit the short musical timing of 50–100 ms will help lift the transient with the ratio controlling the amount of activation 1:1.12 to 1:1.4 that would be a normal range. As with any adjustment to the Sides, avoid collapsing the stereo image in mono. Any increase above +3dB at any point will generally be too much negatively affecting the mono aspect. Aim for a more conservative 1dB lift maximum at points. Using a combination of expansion and static gain increase can help to set an effective width without over exaggerating the Side element. Even a small gain increase dynamically at 0.2 or 0.3dB can be more than enough to lift the Side elements to bring them into line with the mono aspect musically.

Dynamic equalisation

I often find that dynamic EQ comes more into play in an expansion scenario rather than using it to compress. This comes down to the principle of how EQ is used in mastering as discussed in Chapter 13 'Practical approach to equalisation'. Many mastering engineers try to avoid a cut and would rather work around the issues creating positive addition to the sound thus rebalancing the tone across the spectrum. This means when using EQ, it is normally a boost. When using dynamic EQ, the focus is likely to be on boosting and activating the outcome or reducing the boost where a treble addition would be too much.

For example, the perceived top end lacks air/treble, but there are also transient elements that are quite percussive in this frequency area. Application of a static EQ would over inflate these transients. Making this EQ dynamic, setting the threshold to be triggered by those higher aspects creates an 'auto off' to the EQ boost when these louder transient events happen during the music. Set an attack and release to the envelope to reduce in the same way an SC dip in the bass from the kick would be achieved in a mixing scenario. Another approach could be to correct the percussive aspect with a multiband/dynamic EQ and then boost that band. This involves cutting and then addition, not just addition leaving the original transient intact. Also the perceived area lacked treble and was not observed as dull. Positive addition just sounds better in my opinion in every respect, it is less intrusive on the audio with maximum effect. The percussive elements are unchanged and everything else has more air wrapping the treble around these percussive aspects. The Weiss EQ1 in dynamic mode is a really good tool for this. If requiring a tool with more focus on the target, the Sonnox Dynamic EQ has a sidechain EQ shape separate in each band to focus the target quite specifically. Though remember any of these changes must be evaluated at equal loudness to hear the true nature of the dynamic change. There are many processors now available that will create these types of expansion modes, the key as always is testing to be sure how the outcome sounds and not being led by any GUI graphics.

All this said, you could argue, why not just use a multiband? Well that is the whole point of a dynamic EQ. It is not filtered by crossovers, it is an EQ filter shape with the response of an EQ in Q and change of Q relative to boost. It sounds totally different to an area of the frequency range being turned up and down as a multiband would do. It is the shape of the filter that sets all dynamic EQs aside from multiband processors. It is the tool of choice when wanting the sound of EQ with a dynamic aspect. To correct a dynamic issue where something is clearly wrong in a frequency range, a clean multiband is generally more ideal. This difference can be seen in figure 12.1 where the squared flat top created by the aggressive filters on a multiband is compared to a simple dynamic EQ bell shape.

A good example of this would be where a vocal is not really too sibilant, it is just that the rest of the mix is a little dull in the top end. Being able to target where to lift from an EQ perspective, but also have the ability for the EQ to turn off when the sibilance comes forward, where the tonal shift required is achieved without inflation of the sibilance. Normally a very fast attack (0.3ms or less) and release to the sibilance timing would be best. The EQ is dipped around the sibilance of the vocal back to flat so the vocal sibilance is unchanged in tone and everything else is a little brighter. This ties the top end of the mix together and brings the vocal back into cohesion rather than sitting on top at these sibilant moments.

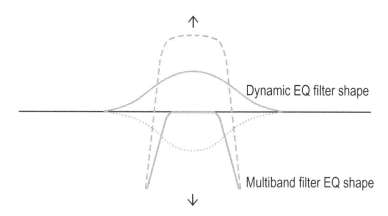

Figure 12.1 Difference in filter shape between multiband and dynamic EQ

Parallel compression

The reason to use parallel compression is to achieve upward compression, which as a named tool does not exist. Upward compression is the outcome of the summative effect of downward compression mixed with the original signal. The method is described as parallel compression. But to make this work effectively in a mastering context, the intent of each aspect within this parallel path must be fully comprehended. The basic principle of upward compression is to increase the RMS in the outcome keeping the peaks intact. This is achieved by creating a downward compressed path reducing the peaks and mixing it back with the original signal untreated. The peaks in the original signal bus mask the overly compressed peaks in the processed signal bus. But equally it needs to be understood that the troughs in the process are not changed because the compressor only affects the peaks. Thus this aspect in both the processed and the original path are not changed. Now summing with a 50/50 mix means there is twice as much of the troughs and less relative peaks; moreover the observation of the trough change is just an amplitude increase. This summative principle can be observed in the exaggerated graphical example figure 12.2.

The use of masking to our advantage is why this type of parallel compression in mastering is often referred to as transparent compression. We only 'hear' the volume change in the signal trough, and not the compression because it is masked by the original signal's transients. In outcome, the difference in amplitude lift would be like riding the fader manually tracking the exact shape of the envelope. This would be nearly impossible to do

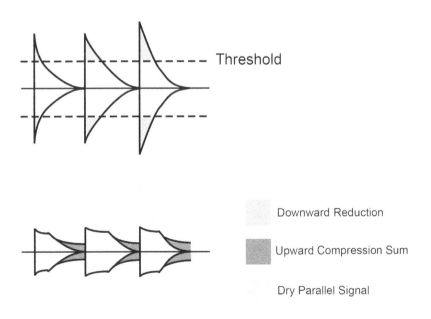

Figure 12.2 Parallel compression change in the envelope shape to achieve upward compression

practically even with automation. Parallel compression in this context works so well because it is the music that creates the tracking for the envelope.

The best tool type to use in investigating this method initially would be as clean as possible. If you have access to the TC electronic MD3 or 4 with the crossovers off (full band), this is a positive tool to engage with, but any clean full band compressor would be advantageous. The default stock plugins in Pro Tools, Nuendo or Reaper DAWs would be effective. The tool just needs to be clean. Thinking back to our first thoughts about how the upward compression is achieved, the compressor is going to reduce the peaks. A slow attack and fast release are not going to reduce the peaks to create the upward effect required. It will just change the shape of the transients. The envelope needs to be properly managed.

A fast bite is essential to reduce the peak detail, but not a limiter or brickwall limiter. This method is discussed later in the chapter. The attack should be as fast as the compressor will go, normally less than 0.5 ms. But it needs to breathe to a certain extent, making less than 0.1 ms often too aggressive and will probably distort the signal. It depends on the selected tool's response in the knee which should not be hard, but rather a semi soft knee. A look ahead control can be useful to make a softer knee compressor trigger earlier in our perception to achieve enough bite in the attack. Any crest control should be set to fully peak as that is the target. Do not think because the outcome is upward the crest should be RMS. The summed outcome is upward affecting the RMS, but to achieve it our compressor must kill the peaks, not destroy or distort them, but really reduce their amplitude.

In practice, once the attack is set the release should respond to the envelope of the music. If it is too short, 'flutter' or 'chatter' (a term more commonly associated with gating) will be audible. As the compression opens back up too fast and as the threshold is deep, the compressor is still triggering cutting in again which creates distortion. If the release is too long the following waveforms dynamic will be 'wiped out' because there is no longer any effective recovery. Avoiding this aspect is easy to do by observing the gain reduction meter recovery. It should be responding to the music, not static in response. In general, a release would be between 200 and 400 ms as a starting point depending on the tempo and pulse of the rhythm.

In fixed beats per minute [bpm] genres, a tempo calculation to ms can be used to set the release in the right area. 60000/bpm will give the quarter note of the music rhythm in ms. For example with a 100 bpm song,

600ms would be the divisible time length. But the intention is to compress each peak and leave the trough, making this full quarter note length too long because the next starts right after our first 600ms. There is no time for recovery. As a theme dividing the calculated ms in two will give the approximate musical hit length to completely kill the transient but allow most of the troughs through untouched by the compression. In this example 300ms would be a good starting point with a fixed 100 bpm. A song at 120 bpm would be 250ms and so on.

This is a helpful guide to get you started. Do not think this is an approach you can use to give you a correct answer. Music is complex, there are often counter rhythms or melodies that juxtapose the main beat and do not fit into this simple structure. But if in doubt it can give you an idea of where to start making it a useful principle to help develop your listening context.

The ratio needs to respond to the signal triggered to deliver enough reduction to effectively reduce the peaks. Generally, the smaller the peak to trough (RMS) ratio, the smaller the ratio. With most mixes, about 3:1 is a good start. In contrast, when using parallel compression on a drum bus during mixing, there is inherently a larger dynamic range in the peak to RMS. The ratio would need to be much higher, 8:1, if not even over the technical limiter ratio of 10:1, basically as much as is needed if listening properly. When mixing, often the parallel effect wants to be heard clearly pulling up to add density and pulse to the beat. But in this mastering scenario, the effect should be transparent. The principle is not to be able to hear the parallel processing, just experience the music sounding better as it has been unmasked in the lower-level detail.

In practice, setting the attack of the compressor very fast 0.3ms, at least sub 1 ms, generally as fast as the compressor will go. The release set to 200/300 ms is a good starting point to 'hear' the compression working with a ratio set to 3:1. The knee should be semi soft, a few dB, though it still wants to be responsive. Now dropping the threshold to achieve trigger and gain reduction, the key to setting a correct threshold level is watching the recovery. In the quietest sections there should be no gain reduction especially in the troughs of the waveform. I cannot over stress how important this is. To truly make 'transparent' upward compression, these sections with no gain reduction when summed back with original are in effect just an amplitude increase with no other change. This is the whole point.

Expect to achieve at least 8dB reduction on the loudest transient parts, if not up to 15dB if the music is very dynamic. This should contrast with full gain recovery back to 0dB. The more this contrast can be achieved, the better and the more effective the overall RMS pull-up will be in the final outcome.

Once this positive threshold is achieved, listen to the release to see where the timing sounds most musical. This will interact with ratio and threshold, and be considered adjusting only one aspect at a time. Remember the original position and make switched/stepped changes to achieve an A/B comparison. Dial in the numbers, think in steps and listen to the difference between the two settings to appraise which is more effective. In general, you will see if the threshold is set effectively having already observed the gain reduction as discussed above. You will be able to find the release time's musical position and adjust the ratio for maximum squash effect. The full contrast in the gain reduction is critical, zero to the maximum possible without negative artefacts. This would normally sit the ratio around 3 or 4:1 but could be higher, probably not lower than 3:1. I can only remember a couple of instances where I have set the ratio lower than 3:1 in this context of processing.

Remember during all this adjustment to only listen to the compressed path outcome of the parallel process, never the paths mixed together. This process path should sound 'over compressed', dull and lifeless when the compression is active, but that is the point. The transient part of the signal has been killed. It should not sound distorted unless this is an intended outcome, in which I would describe this as 'parallel saturation' and not 'parallel compression'. In this scenario you should be listening for a transparent/clean outcome albeit lifeless and dull.

Now it is important to check what is happening with the gain reduction across the track. As a minimum at least one point of the music should be in full recovery. This is likely to be the quietest section. The skill is in adjusting the threshold and ratio to achieve maximum squash on the loudest parts and none on the softest. The compressor is riding the dynamic of the track in the macro even though triggering is targeted directly on the micro. Attention to detail at this stage is crucial to achieve a positive outcome, especially with the more dynamic sections of say vocal or guitar in a breakdown where the spikes are squashed hard and troughs are allowed to breathe openly. It is important that the envelope feels natural even though really squashed at points, it should be musical in timing. The faster the release at this point, the more the general compression will be in sync with the micro dynamic elements, still working effectively with the macro level but bouncing with the rhythm to achieve what will be pull-up between the transients. This can be effective for more consistent pieces of music with a slow pace as the compressor has time to breathe. But on faster material this can often lead to flutter or odd modulation of the amplitude. More than anything it needs to be a timely response from the compression. Though it sounds over compressed, it should be changing consistently with the rhythm and not be working against it.

Finally, the processed signal needs to be blended back with the original clean signal to create the full parallel effect, thus achieving upward compression. Make sure the 'delay compensation' is in place in the DAW or transfer path construction. Removing any delay between the paths is essential to achieve a transparent effect or layer of phase/delay will be added to the signal path as well.

At this point it is useful to understand how a 'mix', 'blend' or 'parallel' controls on compressors is not helpful to achieve an effective parallel mix. If you are going to fully appreciate this technique, you must use a separate bus with a separate volume control and mute/solo, not a percentage or maybe a blend control in dB. These are not helpful as you cannot audition the clean/compressed and parallel outcome back to the original with ease. In short, just using another bus is much easier in opening up other processing possibilities that are not possible with 'all in one' tools or processors.

As a rule of thumb, assuming the dry path fader is at unity, setting the parallel processing path -10dB relative will achieve a positive starting point. Assume the gain reduction on the processed path at its most effective was -15dB, overall this means the loudest transient on the original signal will be 25dB louder than the 'bad compression' that is needed to be masked. Conversely in the sections where the gain reduction is 0dB, the sum will be adding the same audio back in at -10dB to the original. However, if the compressed bus is set to the same unity position, the compression will be clearly audible on the louder section. This is not the point, the bad compression will be clearly heard because the transients of the dry signal are not loud enough to mask the overall compressed signal. To test the blend balance, while playing the loudest section, turn the process volume to a point where there is not an obvious audible increase in amplitude on the louder sections when the process path is muted on/off. This is generally about -8 to -12dB below the original level on the compressed path, but it obviously depends on the programme material. Remember there are no 'presets', each piece of music must be approached with care and attention to detail. If you are unsure of what is meant here by loudness, try bypassing the in/out with the processed path at -5dB and then do the same at -10dB. The one at -5dB clearly jumps in level in comparison.

Perceptively the audio should sound thicker, but not change in amplitude or sound more compressed on the louder sections because it is just the signal's RMS that is stronger. If this is too strong, the 'loudness' effect will come in as there is an obvious increase in RMS and hence loudness. The strong peaks are meant to be masking the 'bad' compression in this loud section. Once this balance has been achieved in the loud sections, do the same in the quiet sections bypassing on/off the process path for a clear 'pull up' in amplitude. The troughs should sound a lot fuller because they are inherently louder. This is true transparent compression because the trough 'pull ups' are the same audio and they are uncompressed on the processed path, and the peak transients that are squashed are masked by the true transient of the original dry signal.

If you are not achieving this contrast in outcome, look back to your settings and go through the steps again. If you do not achieve a reduced dynamic range between the quiet and loud sections, maybe using this processing is not required, so look back to your analysis and contextualise why you thought this upward compression outcome should be applied. If the music is constant in dynamic, maybe downward compression in peak reduction would be more effective. Or parallel saturation is what is required. I will discuss this later in the chapter.

In final comparison back to the original mix, equal loudness comes into play. By summing the two paths, it will be louder overall. To be truly critical, the summed volume of the path needs to be reduced when comparing back to the original. This 'rebalance' is also a useful way to understand how much 'compression' has been achieved overall by the parallel processing. The difference in level is between the pre/post comparison is the amount of 'pull up' achieved overall by the upward compression process. As it is more compressed, the equal loudness is set on the loudest section, and the pull up will be clearly heard when A/B'ing the quiet sections back to the original. If you want to hear the effect without the loudness difference, you will need to relevel on the quiet sections. At either of these levellings, if you make a sum/difference back against the original track, you will hear the upwards compression only. This can be a very useful way to hear the shape of the gain pull up applied to make the envelope as smooth as possible.

Once a clear understanding of the approach to creating transparent upward compression has been achieved, you can start to be more creative. Considering the principle of lifting the troughs, it is possible to negate some of the secondary outcome of increasing the low end by allowing more of the transient to come through. For this to work, the attack time needs to be backed off only a little. If set to 0.3 ms, 0.5ms to 1ms should achieve this, but also make sure our release is not too long, allowing a quick recovery will achieve musical bounce. In other words, the compression will become much more aggressive/brighter. If the timing is correct, a level of excitation to the track, as well as upward compression pulling up the troughs, can be achieved in the louder sections where there is inherently more compression.

An example is a smooth parallel set working effectively at 0.3ms attack and release of 300 ms. To achieve the treble addition and reduce the bass effect halve the release time to 'hear' the difference in process and relax the attack to 1ms. I often use the look ahead delay on the compressor to achieve the correct sync with the attack time. Relaxing this will have the same effect allowing the fast transient to break through, especially when using stepped controls. Often the steps in attack are not finite enough for this requirement, whereas the unit delay is normally in smaller divisions. Overall, the point is to achieve a sense of excitation and density simultaneously from parallel compression without the perception of an overall increase in bass.

At this stage as the processing is working effectively, the release time must be 'in sync' and the mix balance must be correct, doubling or halving the release can change the perspective of the parallel pull up while maintaining the music's sync. When halved, there is more expansion to the pull-up, and when doubled, less bounce. It is often interesting to observe if not always used. These observations all build into our library of listening.

In summary, the outcome of this full/single band process is to achieve dynamic gain reduction with minimum artefact to the audio. That is the primary outcome, the secondary is an increase in the weight of the audio and a fullness and perceptive increase in the lows or maintained a sense of balance in the tone as just described, but it equally still effects the decrease in dynamic range. Any type of parallel compression process will have these primary and secondary outcomes at heart. The more the paths or principles of these initial settings are augmented, the more outcomes are influenced, for instance, by changing the compressor 'tool type', different approaches take over. This is the most obvious change that can be made, by not using a clean and transparent compressor as previously discussed, but one that adds saturation to the tone. This outcome I would refer to as parallel saturation.

Parallel saturation

Changing the tool type from clean to one that exhibits harmonic distortion such as a valve compressor or aggravating the signal with a FET compressor will mean the heavy compressed path will have a significant amount of added tonal colour. If used in moderation it can be a welcome addition to a thin sounding or overly clean mix. Though I feel in principle that this technique is better and more effectively used in the mixing process such as on a drum bus. It can be very positive in bringing life to a mix. The basic difference is in the mix ratio between dry and parallel compressed path. Generally, at least a -20dB difference will be required. Also, the saturated path will be coloured, but less so in the 'non-compressed' sections, thereby changing the feel/tone of the track at these points. I feel this is the problem with this technique. The change in tone between sections often sounds odd to me. But if the music is linear in the macro, it can be an effective approach.

The outcome of the process is adding more treble overall due to the harmonics introduced especially with a valve compressor. I would set an LPF to hem in these harmonics on the compressed path before the sum to avoid dulling the original audio signal path.

Parallel multiband compression/expansion

The aspect to be critically aware of when engaging in any multiband process is the unavoidable fact that adjusting any band against another inherently affects the tonal balance at the crossover points. Hence any multiband dynamics process is inherently also an EQ. If you do not keep this to the front of your mind, it is very easy to be distracted by a perceived improvement in the sound because of the tonal shift and by not necessarily achieving a dynamic improvement. Both aspects should be achieved in all respects, or logically a dynamics or EQ process should be applied.

Generally whichever approach you take, the outcome inherently tends to cause a distinct tonal shift as the parallel aspect is compressing hard relative to the loud sections. If you are moderating the parallel balance correctly, it will create pull-up in the quieter sections. This causes a general shift especially in any weaker tonal areas, but this is one of the great outcomes of this processing tool. In many ways, this can be clearly used to achieve inflation of the weaker sonic areas. The filter band slope needs to be effectively chosen depending on whether you intend to use the process correctively to add excitement/inflation, or a more general shift in balance between bass/treble for instance. Whatever you choose, be aware there will be a strong shift in tone, and you need to be clear what the dynamics are doing alongside this.

A parallel process is there to extend the depth of the troughs between the transients creating that upward lift. The advantage of the banded aspect is it could be used to target an increased depth in the bass while having a minimal effect on treble. Equally this could be the opposite way around or split across the whole spectrum. The impact of a bass lift can have negative consequences in the quieter sections if not well managed, achieving excessive bass pull up at points means automation is needed to control makeup or threshold which immediately makes me question why am I applying this process anyway. I'm not mixing. Remember analysis feeds the principle of processing, not the other way around.

Soft clippers

Almost ubiquitous in modern music, saturation is now a key factor in most digital processes judged to have an analogue feel. In the simplest sense, the digital soft clipper is just modelling the natural saturation/overload of an analogue circuit. This has changed the landscape of mixing and mastering. In the early 2000s and earlier, the need for analogue saturation to enhance a mix during mastering was the theme. As more digital tools have become oriented to have this type of analogue modelled saturation, they have been wilfully applied more and more in the mixing process, which has led to less use of coloured analogue tools to enhance and add tool type enhancement during mastering, as the overall effect of saturation is too much.

At the same time, there has been a new drive for loudness because saturation can create more perceived volume without the negative dulling associated with straight digital limiting. But the trade-off is distortion. If this type of soft saturation/clipping is used too much, the harmonic and distortion generated outcomes becomes obvious. One of the mainstays for soft clipping in a digital path was the TC Electronic System 6000 Brickwall Limiter 2, which has an effective soft clipper and is built in as part of a variety of the 6000's algorithm's paths. This soft clip has a soft through to hard knee and you can drive the level without overloading aggressively. When activating, there is a transition into the clipping, but eventually as the level is increased this becomes hard and obvious. This is one of the general features of soft clippers as opposed to the hard threshold of a standard limiter. Pushing too hard would produce audible distortion artefacts, whereas before this point, you would note less intrusive artefacts in the stereo impression on the Side signal in solo. This is often a good way to tell if a soft clip is running too hot.

A soft clipping effect can be created by using parallel saturation, basically a distortion mixed back in the clean signal. This can be extreme or very subtle by the clean to distortion mix balance and the soft or hardness of the distortion. Experimentation will help you find what sounds effective to your ear. In analogue, this drive in to saturation is a natural part of most circuits, and you will find a variety of responses dependant on their design. As a theme, summing devices or where preamps are involved tend to give the most effective outcomes.

Whatever the method or type, you should be able to achieve about 3dB enhancement on compressed music like rock or EDM in loudness, which is obviously substantial. With more open music in the dynamic, the effect becomes much more obvious as there is less to mask the artefacts in the original source. The transitions in the more open dynamic make the clipping on transients more audible in the micro dynamic.

Another soft clipper is the Sonnox Oxford Inflator process, which while delivering this outcome has much more of an effect on the overall tone than the TC 6000, though with more control to shape. Action is needed to relevel to equal loudness on the output to avoid loudness becoming part of the evaluation in use. In recent years TC Electronic have offered an upgrade to an HD version of the TC Electronic Brickwall Limiter 2 as a software plugin. This has three levels of soft clip which can be useful for different scenarios, but the main 'hard' setting is the most effective, generally for those more compressed sounding genres. Again, a good 3dB can be achieved without a negative outcome, assuming everything else is being managed well. There are quite a few of these tools now available. Ozone makes a similar tool to, but I would use the TC as a mainstay for mastering. The soft modes on the HD can be useful for creating a subtle drive to simulate a valve process in conjunction with a clean tool. The Sonnox Inflator can achieve this with careful use. These softer applications are a positive substitute when the arte-fact of analogue hardware is just a little too much. Any saturator creates harmonics, and it is helpful to hem in any excess, I would always apply a LPF high in the spectrum to soften this at around 28–30kHz or above.

It is also possible to achieve extra perceived loudness by limiting the out-come of the soft clip, adding an LPF high in the spectrum at 34k to hem in. Apply another hard clipper which could take off another dB or so before it is really noticeable, but the sound will be more aggressive in tone. Another limiter can now be applied to smooth the increase out. This final limiter I would suggest is a TP limiter process.

If you intend to make a modern sounding master in the rock/pop/EDM genres, you will need some form of soft clipper to achieve the anticipated level of perceived loudness as part of your palette of dynamic processes. To make sure this processing is positively affecting the sound, I would suggest you use sum/difference to achieve a null to the pre clipper path. The effect of the soft clipping will be heard and in switching back out of the sum/difference and A/B'ing a true equal loudness can be achieved. I would be listening for little change in the tone from the soft clipper, and no negative effect on the transparency, but still observing a reduction in the peak overall looking at the meter in comparison.

Limiting

Where do we start… test, test, test. There are a multitude of limiters on the market that all offer something different in transient shaping from each other. Though they can be split broadly into two types: TP limiters and just plain old digital limiting. A TP limiter will have more response on the same signal because it is using upsampling to increase the sensitivity of the reading at the threshold. In the same way an upsampling meter would show potential overshoots in conversion to analogue on a poor DAC or conversion to lossy audio formats. This is also referred to as the inter-sample peaks to achieve those values by upsampling to 'see' above 0dBFS in how the waveform is really drawn. Aggressive limiting or driving music into hyper-compression will generate TP well above 0dBFS even up to +6dBFS. But it is also com-mon for a normal limiter to overshoot by +3dBFS if using a standard limiting process. In context at 44.1khz the audio is not actually clipping, but it could if the conversion process to analogue is not perfect. Using TP negates this possibility, but it also applies more limiting to control those over shoots. You will hear it more than a standard limiter in comparison for the same peak reduction.

TP has become widely standardised as the generic target outcome for final master output when peak limiting, the headroom of this to 0dBFS is down to your own individual interpretation of best outcome for the music in question. It would be -1dBFS to conform to suggested lossy conversion best practice, -0.3dBFS for CD Audio or -0.1dBFS for loudest digital delivery. Really anything but 0dBFS. For me a limiter process is just an outcome at the end of the chain to skim off any peaks on a good sounding master or to level peaks earlier in the chain to help a post dynamic process to have a more effective trigger having imposed a range on the audio. Either way, the amount of reduction I would apply is minimal, and certainly under a couple of dB reduction at any time. For me, loudness comes from good frequency balance and control of the dynamic range, not limiting. The focus should be on the RMS and not the peak. Get that right and the peak will be easily managed no matter what the limiter used. But as a final process I would always use a TP limiter. Having extensively tested the TC Electronic, Sonnox, Flux and Weiss offerings as well as many more, I would use the Weiss in a T2 mode setting in preference every time for this final TP function, assuming the audio has already been restricted with a normal limiter or clipper. It is more transparent and accurate in my opinion, but my tests always continue as tools develop. In normal limiting early in the chain, anything goes, so I often use the UAD Precision Limiter because it is a low DSP resource and can control the peak with little artefact with a low gain reduction 2dB or less. I expand on the use and functional setting for limiters and soft clippers in Chapter 14 'Navigating loudness'.

Parallel limiting

Bringing our thoughts back to parallel processing. Utilising these limiting principles set out above but adding a parallel path to blend a positive upward compression outcome can be achieved transparently though only with a marginal pull-up in level. More aggressive application of the limiter tends to show excessive tonal change in the outcome because of the dulling from the limiter, making the tonal swing quite noticeable. I personally have always found a more transparent outcome with full control of the attack, release and ratio with a compressor, meaning the parallel compression track can breathe more. But it is another option to experiment with in testing, and some engineers find it very effective.

Summary

You need to explore the principles of dynamic application and find solutions that sound effective to your ear, and find a combination of processing that leads to a soundscape which is appealing to you. No one can tell you how to do it best for your outcomes. People can guide you with the principles and methodology, but you need to be the listener and the judge to find what sounds best to you. Remember there is not a correct answer, but there is one that can be justified. If you cannot, then you need to go back to the beginning and evaluate your approach again.

Practical approach to equalisation

To boost or to cut

As a theme, I would work to apply a boost rather than cut. If a mix is too bright, boosting the lows and softening the transients with downward compression is more effective than a high cut. The same goes for too much bass; lifting the high is often better than cutting the bass unless this shift is a noticeable amount (over 2dB), then maybe a cut in the low and boost in the highs would 'tilt' across the audio spectrum to achieve an ideal 'flat' outcome. The benefit of this approach is it is more active. The audio sounds enhanced rather than pacified.

A straightforward observation of any mix in principle would be that everything in it was intentional and signed off by the client. That does not mean its tone overall is correct, though it does direct our thought towards not removing content but controlling what is there. When observing in analysis and contextualising a processing action, it is important to think positively. For example, if the track had too much weight in the low mids, adding bass and treble to balance out the excess perceived mid-range would be a positive action. If the track was overly focused on the bass, simply add treble. This is a positive thought and not a reductive cut. The other aspect is phase distortion – a boost always has significantly more phase distortion than a cut, therefore it will always sound more active in contrast unless using a linear phase EQ. You could argue that cutting must sound more passive, it does, but why not just use a linear phase EQ instead and boost.

The same principle applies about not cutting smaller areas of sound unless absolutely necessary. Even if there is a lump in the low end at say 150Hz, feathering (discussed later in this chapter) boosts around this at 110Hz and

DOI: 10.4324/9781003329251-14

190Hz masks the lump creating a wider lump but its focus is more defused, boosting the mids/treble above 250Hz, and bass below 50Hz balances this bigger lump created out. No cut is needed. This active approach sounds fuller because cutting always creates a ripple across the rest of the spectrum as much as boosting can generate harmonics. In the end, there is always a time when a cut will be the only option, though it is just about changing perspective. Start to think positively about how to embrace the use of mastering EQ.

There is not a correct way to do anything in principle, but any approach's effectiveness can be debated. For example, if engaged in engineering for live sound, boosting would activate the harmonic content and would be more than likely to achieve acoustic feedback. It is better to actively cut the opposing aspect to achieve the same shape, but without amplitude increase and phase shift. Rebalance the level overall with the channel fade, again passive.

A negative equalisation approach in mastering often leads to an outcome that lacks lustre especially in the transient detail. The longer wavelengths, the more troublesome it becomes to work positively, and sometimes there is no choice but to cut. Though in our own minds, being clear about why a filter needs to be applied and trying to positively work around the negative aspect is an effective approach. Often the problem area can be easily masked by raising the sections around it, rather than cutting the problem itself. Shifting the spectrum around this issue to rebalance to 'flat' means nothing has been taken away from the original music, just different aspects of the music activated. This principle in approach is discussed further in section ix. 'Extending the frequency range'.

Shelf or a bell

Is the target an individual instrument/part or is the purpose to control the whole tone? If the latter, a straightforward shelf will give the required tilt in perspective. In looking to affect a specific instrument or area, for example, brighten the vocals or control dull piano mids, a bell shape would be valid. What should also be kept in mind is that a standard shelf sounds less active than the bell shape. A bell changes amplitude across the whole range of its shift, whereas a shelf by the nature of its name reaches a plateau. It is perceived as an amplitude change across this levelled section, which means it sounds more neutral without a focus on a centre frequency. This is discussed in Chapter 7 'Equalisation and tonal balance'. To activate an area, looking at bell shapes to introduce focus in the centre frequency is preferential. Some would argue you can do this effectively with a Gerzon resonant element on the shelf. This is true, though personally I favour keeping a smoother shelf shape and using the bells to activate frequency points to create a more accurate outcome with a good sounding mix. Though this can increase the amount of phase shift relative unless using a linear phase EQ. I have found Gerzon shelf shapes are effective where a big change in tone is needed to help sculpt an outcome, and are more useful in restoration where tonal shifts can be large.

Pre or Post

Positioning of the EQ is critical, and there are two things to consider: for practicality, the equipment available, and secondly, the state of the mix in terms of the need or lack of need to apply corrective processing. In principle if a cut is needed, it means information (music) is to be taken away. Dynamics processes should not be triggered by a signal that has to be removed, hence the EQ would go first in this instance. This would suggest if there was only one outboard EQ all our changes would be applied first, before compression. But there is also the dynamics to consider. If the dynamic range is large, the louder parts will in effect have more EQ than the quieter parts, this is especially true when boosting aspects. As the dynamic range is compressed, the original dynamics will be imprinted in the tonal change even though the dynamic range is now controlled in amplitude. This can be helpful when wanting to maintain the bounce of the original dynamic of the music. But equally, it can make overly dynamic material sound uncompressed and disjointed, even after processing correctly with dynamics. In this case the EQ boosts are better placed after the main dynamic processing, and cuts before, meaning in a purely analogue transfer path, practically two separate EQ units would be needed.

Simplifying this thought process to an aesthetic one, if the music needs to be activated, the EQ should go first pushing into the dynamic processing. If a smooth change in tone is required it is best to EQ post dynamics. Both observations assume any EQ cuts that remove energy happen before dynamics to maintain the most effective trigger. But as previously discussed, working to avoid cuts should be a theme in the first place using positive EQ. This would mean these are likely to happen before the compression to have a more effective tonal balance going into the compressor. This also makes the assumption that if the music does not sound close to a correct tonal balance in the first place, the EQ is going to go at the start to sculpt the tone into a rational state to compress, even if this is all boosts. If the mix is noticeably incorrect in tonal balance, the dynamic processing and tool type are not going to 'fix' the outcome. It is important to get the tone generally in the correct zone and then apply dynamic control. In short, there is no point engaging in dynamic manipulation if the source is tonally incorrect, in the same way as a cut in tone should be before the dynamics. There is no point compressing something that will then be changed. If there is a positive tonal balance in the first place, the EQ will be after the dynamics for the smoothest response and minimum intervention on the audio. But practically, on the majority of masters I would probably engage in a bit of both, pre and post before to get the tone in the right area without too much activation, and after to highlight musical accents such as the rhythm or voice.

Relationship between filter frequency and gain

Previously I pointed out that the notion that stepped frequencies restrict our ability to target audio is misguided, as there are a multitude of benefits to step controls. The fact that the frequency and gain are fixed positions does not mean the gain has to be changed to affect the perspective of amplitude, the filter position can also be adjusted.

When EQ'ing a sound, the aspect in focus to be changed is in a specific frequency area, even if this is across the whole bass end or treble. For example, apply LS boost in the bass with an HPF added by default to avoid lifting sub energy. In auditioning, if our boost sounds too much, reducing a gain step sounds too little. Leaving the gain at the overactive position and adjusting the frequency of the shelf to drop a step lower means the area of boost is reduced because there is less frequency range to activate and hence less gain is applied overall. You might even find increasing the gain a step up at this point achieves the optimum outcome. Just because the gain or frequency needs to be adjusted, does not mean that the specific control needs to be used to affect the outcome.

I feel this is an important aspect of comprehending how to engage with mastering EQ. The point is to create a shape that counters the current musical imbalance. There is a need to actively use frequency and gain to find the correct balance. The quality factor (Q) can help with the scope of each filter, but generally the creation of focus with the Q is best avoided, otherwise it will sound like EQ has been applied – never a good outcome on a master. It is not positive to hear the EQ shape directly in my opinion. This is a very different perspective from mixing where creatively anything goes and often to amplify or drive a sound musically forward the resonance from the EQ wants to be heard.

The premise of moving the frequency point to affect the gain is valid just as much with a bell as with a shelf. Moving focus away from the sound to be corrected can have the benefit of smearing the frequency range of the instrument, making it wider than its original context. Another way to describe this would be feathering the frequency.

Feathering frequencies

A classic outcome of feathering a frequency is where the focus of the original EQ boost has been smeared by the use of surrounding additional boosts to distract the listener's focus. When the analogue domain was the only processing path available, this method was very helpful indeed to soften the effect of an amplitude boost by masking the centre frequency of the gain change. For example, if a bell boost of 2dB at 100Hz was applied, the feathers would be applied as a bell boost at 80Hz and 120Hz around the original centre. The gain energy would now be redistributed by halving the centre frequency to 1dB and adding the removed gain to the two feather frequencies at 0.5dB each. Total gain is still the same as the original boost, but the focus is distributed across a broader range of frequency, meaning the centre frequency is less obvious and more masked. This is not the same as increasing the Q on the original EQ where the centre focus will be maintained, the smear in focus happens by the two feathers defusing the centre boost, making it softer in tone. This outcome can be observed in figure 13.1. As a theme, the original Q from the original frequency would be kept and used with the feathers, but there is no fixed ratio to the frequency feather. They normally work best if about equal in distance from the original frequency set. The higher the frequency, the wider the numeric, and the lower the frequency, the smaller this numeric distance will be as the wavelength is

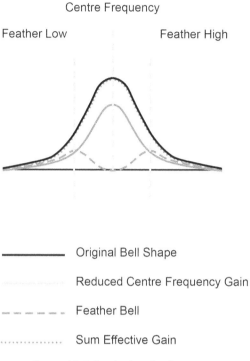

Centre Frequency

Feather Low   Feather High

——————— Original Bell Shape

Reduced Centre Frequency Gain

Feather Bell

Sum Effective Gain

Figure 13.1 Feathering the frequency

longer. It is the smear around the centre frequency that softens the effect of the EQ, hence the feathers need to be in proximity to it or it will just sound like three separate EQ boosts. Avoiding too much separation is straightforward on a stepped EQ as the range of selection is restricted.

The root of the need to feather in analogue is because the resonance of the bell boost is emphasised by the phase distortion in the EQ. When cutting there is minimal phase shift because a feather should be used when boosting, it is ineffective when cutting. In a modern context and with linear phase EQ, if a transparent shift in tone is required it can be easily achieved. But the premise of blurring the centre frequency is one that can still be used to soften a lift in gain while achieving the positive tool type outcomes from the phase distortion without its potential negative effect in over-focusing an area. For example, in wanting to pull the guitar clarity out of the Side of the mix, you could apply a boost at 5kHz with an analogue style EQ tool type to 'add' activation to its intent, but you can add another lift at 5.5kHz to soften the distribution of the resonate focus of the bell across the two filters. This single feather blurs the focus of the original bell boost. There is not a correct ratio to calculate each feather used. The decision is around how to distribute the energy, meaning you need to be aware of the potential relationship of each feather. With a step EQ the choices are obviously fixed.

This type of single feather is often useful to blur the corner frequency of a strong LS lift. It can help to move the focus of the shelf corner frequency at the top of the boost. This creates a half Gerzon style shape, but sounds more diffused than a Gerzon resonant HS and is easier to target in frequency range.

Hearing the gain is not the same as setting the gain

It is easy to apply too much EQ. When EQ'ing, an engineer is listening for the change to observe when the change has been applied across the intended frequency range. It needs to be clearly heard. This is true in mastering as much as mixing, but in mastering, it is best not to hear the EQ itself, just the effect. After setting a bell or shelf, back the gain boost off from the clearly audible position. This is definitely the case for bell boosts being the most active. I would suggest halving the applied boost when you feel it is in the correct frequency position and re-evaluate, bypassing the band on/off to 'hear' the difference applied rather than listening for the change while the EQ is active. The change that is already there in the boost often masks the effect of the larger gain change. Again, step control or method helps to achieve a clearer observation of the dB shift and manage equal loudness. Equal loudness is the key method to audition a change in tonal balance to avoid over boosting. The more this becomes a habit, the easier it is to find the correct gain range in the first place because you are becoming more critical as to how powerful EQ is at sculpting the sonics of music. EQ is very easy to overuse. For example, highlighting the vocal diction in a mix by adding an EQ to the Mid path around 8kHz, Q2. This should be applied dynamically if the vocal is not present for large sections of the mix, remembering the principles of momentary or static. Increasing the gain to clearly hearing the EQ, the lift may have to be a couple of dB. Once set, halving the boost to 1dB and auditioning on/off, the vocal should 'unmask' but not sound EQ'd. Setting the gain back to where it was original it would sound like EQ has been applied to that area. You may find only a minimal amount of 0.5dB has the desired effect, remember we are listening for the vocal diction unmasking, not to 'hearing' any change in frequency balance. The amount of EQ required to hear 'it' is not the same as the lesser amount to hear its effect on the music.

MS elliptical EQ

One of the most powerful applications of MS with equalisation is one of the simplest. Encode from LR to MS, apply an HPF on the S path and decode back to LR. This enables the bass end/sub to be made truly mono without affecting the rest of the music. This pathing can be viewed in figure 13.2. Mastering engineers have been using this function from the advent of stereo because cutting a lacquer with too much energy (bass) in the Side and the vertical motion of the cutting head will cause lift off or break through during the cut. Elliptical EQ is the safety filter in the cutting transfer path to avoid excess energy in the stereo impression.

In a modern context, this allows music that is meant to have a strong bass focus to be absolutely mono in the lower frequency and especially sub. The

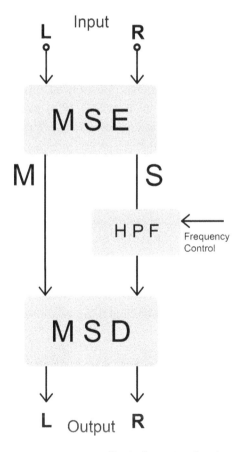

Figure 13.2 MS Elliptical EQ signal path

lateral motion of the cutting head is mono. Though a lot more caution is required with more spacious music, as not to take away the positive width and effect of the acoustic room sounds in the bass end. This can make a master sound unrealistic. But a high pass at 30/40Hz can help to make the centre feel a little deeper and focused in the mono sub relative to the Sides. Remember, modern mixing and mastering is mainly for digital formats that do not have a restriction in the energy remit of the Side element. If the cutting engineer thinks the energy needs reduction, they can apply it if the digital master is to be cut to vinyl.

As a theme, a standard elliptical EQ will have a range from 20–300Hz and a soft roll off of around 6dB/Oct, musical, to avoid hacking the bass end out of the Side element. I generally prefer a Bessel filter 2nd order to hold in the lower aspects which avoids removing any of the 'good' energy from the Side. But different songs sometimes require different responses. Correct application can help to move the sub forward in perspective and tighten the lows. Some engineers have referred to this as visualising an upside-down triangle in the lower frequencies with the point focused on mono at the lowest frequency point. The image widens out as the frequency lifts. The narrower the triangle, the less width in the bass and so on. Positive critical evaluation in A/B should be used to avoid removing too much. Conversely a Dubstep tune could have a very high filter above the low mids at 500–1k to really focus the whole mids/bass end to mono and avoid any bleed. It is highly likely when being played in a club that the lower mids, bass and subs are mono in playback. This high filter will give the master true phase focus in the whole low end.

There is nothing magical about the physical version of this circuit other than it can be used in your analogue path. I always prefer to tailor my own in digital, pre any analogue processing. Remember when removing information, it should be close to the start of the signal path, not after the event. Though it may be positive to apply again after an analogue return in the digital post path to make it truly mono; any bleed from the analogue path processing is also removed. But when stacking the same filter, the effect can double the roll off overall. It's something to be aware of.

MS practical application

The ability to change the impression of the frequency range across the stereo impression is a powerful tool to unmask. An instrument in the mono aspect can be helped to move away from elements in the Side or unmasking both.

This is not always about correcting frequency balance, but readdressing registers across the stereo image. In the same way as in a mix, cutting into the bass a dB at the kick fundamental gives a little more space for each. The same principle of pocketing the frequency across the stereo image can be achieved. For example, if focusing on the higher edge of the vocal at 8kHz in the mono was needed to bring the vocal forward in the mix, cutting that same frequency in the Side could be an option. But thinking about our positive EQ mantra, lifting the area above or below at 10kHz or 6kHz to move the energy away from the 8kHz position will give space and excitement between the Mid and Side, making the imaging wider and unmasked. The same could be applied in the low mids, lifting the energy at 250Hz in the Sides. A little at 1–3kHz in the Mid would push the guitars or synths in the Side deeper and give a little more bite so it would sound like the 500Hz area is reduced overall. Create space in the centre to open this range in the mono aspect. Just spread the weight of the energy from either the Mid or the Side to give a more defined stereo impression.

This pocketing can be directly matched in the lower frequencies to redistribute energy. If the vocals are a little full, cutting at 250–300Hz in the mid and boosting the same frequency and amount in the Side can often open out the vocal area without it sounding like the cut has 'thinned' the musical context. Remember, this is about redistribution of energy and not fixing the tonal balance in overview making a linear phase EQ the ideal tool.

Extending the frequency range

When mixing, it is important to sit different aspects in their correct frequency ranges. This often means instrument tracks are filtered and correctly hemmed into their ideal frequency range. Hence the instruments are unmasked and well separated. From a mastering context, it means there is no need to apply filters because this has been correctly addressed in the mix. But this does not mean all the potential frequency range available is being utilised. There is often a need to extend the range of the frequencies towards the extremities of the resolution during mastering to add density and definition. This is most clearly evident in low frequencies, where pulling the bass/sub deeper with an LS at 25 to 30Hz and an associated HPF at 12 to 18Hz. The lows can be pulled up as much as needed to lower the frequency range distribution, adding power and clarity in the subs/bass. A Gerzon resonant element can help to push the sub without lifting the bass too much, as the active part of the shelf is steeper. Though I prefer to shape the end of the shelf rise with a bell to lower it, but importantly without cutting the original frequency range overall. It is just there to sculpt and hold the shelf shape in.

The same applies in the high top end. Lifting above the audible spectrum can give a smooth extension of the highs to add energy. This HS could be 24k upward, again hemmed in with an LPF at 30–40kHz. It could also be achieved with a bell boost at 28kHz or around this area. This is sometimes referred to on some EQ as an 'air band' because it lifts the harmonics, and you can hear the energy change through the audible spectrum. The Q for this would normally be around 2, while a Q of 1 would be too broad, clearly reaching into the audible spectrum.

This same thought about frequency distribution is often required with a less honed mix, where the kick has too much focus. There is a need to redistribute the energy across a broader range of frequency, otherwise the intensity of the kick will prevent the dynamic restriction overall. Achieving an appropriate density without distortion in the low end would be impossible without redistribution of the kick focus. To approach this, find the fundamental frequency of the kick and apply a broad bell cut across the range of the kick distribution. This would be a Q of 1 to 2 to give a broad enough focus. For example, if the kick centre was at 70Hz, a bell cut set at -3dB Q2 would reduce its frequency energy. Applying an LS at 35Hz to lift the sub and pull the kick frequency lower might require a boost of 5–10dB depending on how poor the original kick frequency distribution is. Add a high pass filter to hold in the lowest frequency area of the shelf lift at 12 to 18Hz to avoid unnecessary RMS energy being added. This all pulls the focus of the kick away from its centre frequency. If you have achieved this correctly, you would see a noticeable drop in the RMS energy on the meter in comparison at equal loudness, yet the kick will sound fuller, deeper and more impactful when listening. This sculpting of the frequency range can be observed in figure 13.3. Broadly this can be observed as a type of feathering and as the positive active EQ approach because the aim is not to cut but redistribute the mix energy. There is less focus and more solidity or 'fatness' to the lows.

The same logic can be applied to the high end where the top end of the music is overly focused in an area. For example, this could be 6kHz around the vocal top end, synths and guitar air. Again, a broad cut with a Q1 at 6kHz and an HS at 18kHz and with LPF holding it in at maybe 30kHz. This would soften the harsh frequency area and re-distribute the energy over

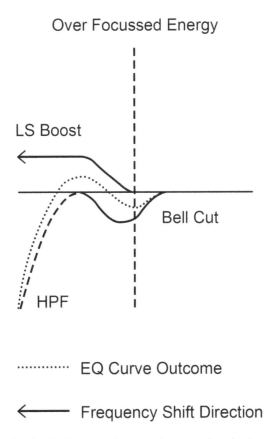

Over Focussed Energy

LS Boost

Bell Cut

HPF

·············· EQ Curve Outcome

⟵——— Frequency Shift Direction

Figure 13.3 Redistributing a mix's energy by extending the frequency range

a broader spectrum. More clarity with less harshness is what to listen for, spreading the frequency distribution.

The aesthetic of tool type

I think all engineers develop favourite EQs as they deliver on outcomes to their ear. The more they are used, the easier it becomes to really appreciate the refinement enjoyed from utilising it. This is driven by the correct selection of tool type for the requirement trying to be adjusted, but it is also subjective to our own likes and dislikes. For example, I really enjoy the sound of the LS and HPF combination on an EMI TG12345 curve bender. The LS bass smear with the phase distortion adds real depth when required. But its other filters are not my first choice, though the high mid bell boosts can bring clarity to a muddy vocal in its phase focus around 8khz. A Weiss EQ1 has a musical outcome with bell filters in minimum phase mode dynamically, but I prefer the linear phase outcomes with EQ116 on Sequoia, which just moves amplitude around in a way that works for me. The TC electronic system 6000 MD3 or 4 EQs sound musical, though the new native plugin versions only go up to 20kHz in the top end which restricts their use. The shelfs on the original at 3dB Oct are brilliant for tilting the frequency range. But another mastering engineer may dislike all those outcomes. It really is subjective. But it is impossible to ignore that the linear phase is really clean and boring as opposed to a valve EQ which will add excitement, harmonics and saturation. In between these two extremes are many different variants due to the circuit design. All this focuses down to the principles of test, test, test. Know your equipment and you will know what outcome will be appropriate for your perspective.

The negative effect of EQ

Why do we EQ? A positive answer would be to make the frequency balance sound more coherent, or to make different instruments fit together. But I feel the elementary answer in mastering is to unmask. Sometimes I feel it can be easy to get lost in the application of EQ unless sticking to our initial analysis. Often our observations in overview are straightforward. For example, noting the mix is dull, meaning it only requires a straightforward response, an HS at 3k to lift and an LPF at 24k to hem in. In contextualising this link, apply your focus on the task at hand to correct the dull observation. Because tools are available does not mean they should be used. Staying on top of our want to 'fix' the frequency range in real time variably and keeping our focus on application of the evaluations from analysis will lead to a better end result. Our outcomes are less likely to be intrusive to the general sonics of the music and its original intent. As I have said previously, this is something I have grown out of and in doing so have become more respectful of the power of EQ to change the context of music.

Notches

I never use them in a modern mastering scenario because they are too intrusive on the musical context. But there is always bound to be some scenario, so never say never! Anytime holes are cut into complex programme material, ripples are created throughout the context of the frequency balance in the same way as boosting creates harmonics. In tackling one error, the outcome is a thinning, making more errors ripple up the spectrum in areas that did not need any processing. In my opinion if the mix is this wrong, a conversation is needed with the mix engineer to address. If they cannot fix it, it is better to ask for a full stem breakdown. Finish the mix, get it signed off by all involved and then master it.

The only exception to using a notch would be in restoration mastering where removing hum or buzz to rectify a really poor source as this is often the only version that exists. But this is about fixing a bad source. In restoration it is clearly better to hear the music than not at all. This is not the same scenario as improving what should be a very positive mix to master in the first place.

Summary

Context is key in any EQ scenario. This brings us back around to analysis and evaluation. When spending time in analysis observing what needs to be enhanced, correct tool selection and approach can be applied to add impact and clarity to the mix. This often means EQ is a secondary element of the chain to the dynamics, as they will affect tonal change in processing. Comprehending this means you will use EQ where needed and not as a 'go to' to change frequency balance.

Always remember, just changing the overall amplitude changes our perception of the frequency balance. Clearly, there are more ways to enact rebalancing of the tone than linking any analysis of frequency made to EQ.

It is not unusual to master a track and apply no EQ, or just pass filters as the tool type and secondary outcome of the dynamics enable effective change of the tone. EQ can be a blunt weapon in the wrong hands. With knowledge of dynamic processes applied, it becomes an artful tool to enhance any music's directive where needed.

Navigating loudness

One of the questions that I am asked more than anything else as a mastering engineer by other engineers and artists is 'what is the correct output or target level for a master?' There is a straightforward answer: There isn't one. But there is a caveat and context to help assess what is appropriate – the mix.

Realistic approaches to evaluate loudness normalisation

Every mix is a unique piece of music with its own inherent dynamic and potential effective loudness. If normalised to a fixed or suggested level, whether it be peak or average, its perceived loudness will be relative to that internal density and dynamic range. In current times, this could be using ITU 1770 [9]/EBU 128 [10] or another LUFS or a dBr scaling. These aspects will change in the future as newer 'technologies' for measuring loudness become fashionable. That is progress to some, but one aspect that will not change is how the master reacts in normalisation against its mix source. In any comparison, the master should always sound better than the mix under each and every levelling circumstance. Whether the focus is on peak, TP, RMS or LUFS, if the master does not sound better than the mix, why are we mastering it!

I should also point out the actual true comparative in all audio levelling is equal loudness. When we use our hearing to critically assess loudness differences and not rely on a measured scale. At this point in audio analysis, technology has still not made any function or tool that can measure actual musical loudness in the same way that audio engineers can easily do by listening.

 DOI: 10.4324/9781003329251-15

This is clearly borne out in any cross-genre loudness levelling with an RMS or LUFS meter. A folk song, for example, does not sound 'levelled' to a rock song and so on. There will be a discrepancy in apparent equal loudness level. An ITU LUFS meter will be more accurate than an incomplex RMS meter, as the LUFS reading is weighted, but neither will be correct to our perception of loudness. If you are interested in how these metering systems technically calculate loudness, I suggest reading the materials available on the EBU and ITU websites. These are very informative.

An ITU or EBU meter is designed for all audio, post-production outcomes in TV, radio or film. That is why they are an ITU/EBU standard [9], [10]. They suggest one of the main rationales for the development of this measurement was *'that listeners desire the subjective loudness of audio programmes to be uniform for different sources and programme types'.* [9] I agree 100% with this statement. As a consumer myself, I have no desire while watching TV to have to turn the volume up and down every time the adverts come on. Levelling is a very good thing. Though context of this is important, the key is in their respective names: 'T' for television and 'B' for broadcast in terms of the intended arenas of use for this type of metering. There isn't a 'music' meter standard, because who is to tell an artist how compressed their audio should sound? No one, it is their art. On the other hand, a TV programme requires a blend of music, dialogue and Foley. This delivery changes the context of the use of music into a new complex dynamic range over the programme time. These mixed sources make it more troublesome to navigate relative loudness. That is what an ITU/EBU meter is designed to measure, and it does this well. Apple Music and Spotify and others have adopted these standards not because it is the correct thing for music, but their distribution models are a varied and complex set of outcomes. In their consumer delivery content, there are audio books, podcasts, music, live radio stations and the list goes on.

This means the adoptee's standard works well in levelling these disparate outcomes for the consumer. But it is not accurate in the way a mastering engineer would observe the sonics, and certainly not with music. It is just somewhere nearer to being more levelled though generally better than not being levelled at all. Which from a consumer perspective makes the function quite useful to them. I am not 'anti' this or future methods of making the life of the consumer easier. It is just important to maintain a realistic relationship in use when observing these automatic levelling outcomes.

This broadly adopted standard is not the law for music delivery, or the correct outcome. To suggest it is, would be very dismissive of the music source and the artist's potential intent. The focus should be on making sure the outcome is being enhanced from mix to master and not chasing some comparative generic levelling. Though every outcome has an element of comparison in how it will be digested by some consumers, it will not be all users by any means. Even then, with some of these platforms the consumer has to turn the automatic levelling system on. In this way many consumers are still listening in a peak level comparative anyway. But if you load a song on YouTube, it will be levelled on ingestion. This term 'ingestion' is used to describe the act of loading a media file or set of files to a host. In this case, the master should sound louder than the mix version, otherwise the louder mix version would sound better regardless of the mastering that had taken place.

This is one of the most straightforward evaluations to make when comparing outcomes. If the mix sounds better than the master under level normalisation, it is because of the perception that louder sounds better. Although this is probably a fault in the master being overly compressed and/ or contains too much density in the lower frequency ranges making its tone imbalanced.

Level normalisation is not something to be worried about. What needs to be understood is where and when it happens. What we should be asking is:

What does it change in the perspective of the audio under comparison?
How does the consumer use it or abuse it?
What does this mean for the dynamic range potential at a master's output?

Also bearing in mind the need to be aware of peak level comparative in loudness at the same time. Level normalisation is not the norm, it is just another potential outcome in consumer use as much as an offline download, CD, cassette tape or vinyl playback would be.

All this can seem to be a lot to take on board and manage, but a positive approach to tonal and dynamic balance in the first place works no matter what the output. Audio that is made flat in tone balance and levelled will translate effectively for all scenarios, both for past and present and future. The rules of focus around tonal and dynamic balance do not change with the format as such. The format adds potential restrictions, but the principle of an even frequency balance has always and will always sound best.

Streaming platforms and their loudness normalisation systems

In 2023 Spotify premium (lossy audio) has three levels of consumer play-back under normalisation: −19dB, −14dB and −11dB LUFS. The consumer gets to choose the level of 'loudness' they want depending on their listening preferences, if they are even aware of what normalisation is. But by default normalisation is currently on and set to -14dB LUFS on most apps. This potential flexibility in playback levelling is not a bad thing in principle. It is convenient for the consumer and is probably a good thing on balance unlike lossy audio delivery. Spotify premium is still lossy, which though helpful from a data perspective is not a positive thing for the sonic quality. Though they are suggesting CD quality is on its way; hopefully when you are reading this lossless will be the norm.

From a mastering point of view, the first thing to understand is that the platform does not normalise on ingestion, hence the original peak and RMS ratios to 0dBFS are maintained as they are inputted into the system. The normalisation happens on the app selected for playback, some of the more simplistic playback apps such as smart televisions do not have a normalisa-tion outcome at all. The observed comparison between different songs and albums by a consumer would always be at peak level. But if a premium customer with normalisation on selected the 'loud' setting of −11dB LUFS in the app on their iPhone, any music played would be normalised to that LUFS average. If our master was levelled and ingested as suggested by Spotify at −14dB LUFS with a peak of −1dBFS TP, when this master is played back with a potential dynamic range of 13dB levelled to −11dB LUFS, the peak infor-mation could potentially clip by +2dB at 0dBFS, remember that 1 LU = 1dB. Luckily, the engineers at Spotify thought about installing a limit function so this master would now be limited by 3dB as Spotify's limiter is set to engage at −1dB, with a 5 ms attack time and a 100 ms decay time. This is not the best outcome for dynamic range control of our mastered audio. If however the master was optimised to a dynamic range of −11dB LUFS or lower with a TP of -1dBFS, the normalisation would just leave the song levelled at that −11dB LUFS level. If mastered to -10dB LUFS it would be levelled down −1dB and so on. But importantly no sonic changes to the audio would be made except for its conversion to lossy. In normal playback mode, the track mastered to −11dB LUFS would be turned down 3dB to −14dB LUFS and in quiet mode 8dB down to −19dB LUFS. The sound of the mastering has not changed, it has just been reduced in amplitude overall. But importantly, these more compressed dynamic ranges when level normalisation is off sound significantly better than anything ingested at −14dB LUFS just because they are louder in peak comparative. As we do know, louder really does sound better.

It is worth noting again at this point Spotfiy currently has level normalisation on by default in premium, to change relies on the customer knowing this is something they might want to do. Equally they have to find where it might be located in the software and have the time and inclination to action it. As discussed in Chapter 16 'Restrictions of delivery formats', convenience often trumps any improvement in perceived quality. In the loudness normalisation context, a lot of users will just not know or comprehend what it is, making this comparative their impression of the audio. This is unless a third-party device such as a TV or smart speaker is being used, in which case the opposite will be true. Either way around as mastering engineers we have to be aware of all the potential outcomes and respond to every single one to achieve an optimum master which will work both now and in the future. In contrast to premium, listening to Spotify free in the browser has normalisation unavailable and there is no way to adjust this. In the desktop app this can be turned on or off. But again, any of these aspects could change tomorrow. Your master just has to sound great in all these potential outcomes.

On Apple Music, the consumer 'Sound Check' button levels all audio to −16dB LUFS, a simpler principle which is currently off by default. But it is best to make sure our LUFS does not have a level lower than this to avoid any negative effect on peak level relative. As Apple says they only normalise as far as the peak will allow with sound check on. This means that more dynamic music, often wider in dynamic than −16dB LUFS, will not be increased in relative level. This is still very quiet in any peak based comparison and is quieter than the average with Sound Check on. Some engineers are minded to submit music levelled to the media broadcast standard of −23dB LUFS. This is madness if you actually test what happens to the artist's audio in the varied possibilities of consumer use, especially comparative to other material as they are bound to do.

Apple Music, Amazon and Spotify platforms allow the consumer to turn on and off loudness features if available in a given app. Some other platforms' normalisation is always on. All these specifications could change at the whim of the platform, as they have done in the past. But in adherence to the principle that the master has to sound better than the mix in peak, RMS and LUFS normalisation, the mastered outcome will be positive in quality for both now and in the future. Of course, it goes without saying the master should always sound better at equal loudness in comparison to the mix, or we should really question why we are mastering at all!

Limiting for loudness

Alongside the principle of achieving effective loudness outcomes in level normalisation systems, there is the age-old comparative level issue. The loudest output in RMS-to-peak ratio tends to sound best even if it is hyper-compressed, and the quieter output sounds dull and lacks life in comparison. I personally would avoid making this type of comparison, something which has been pushed too hard is just that. It is not constructive to 'chase' an outcome that makes our audio sound worse for a competitive evaluation at 0dBFS against a master that fundamentally sounds hyper-compressed – referencing the bad to make more bad outcomes.

The opposite will be true if the audio is level normalised against the hyper-compression outcome. The wider dynamic range master will sound brighter, fuller and louder. What should be observed and referenced are songs that sound 'loud' in all scenarios as they must be delivering positive outcomes in songwriting, production, mixing and mastering to achieve those results. Taking these lessons on board is critical to achieve sonically superlative outcomes.

To achieve a 'loud' peak to RMS, it is prudent to look towards another form of dynamic restriction to deliver an effective result – soft clipping. Application of a limit stage will only gain a couple of dB maximum before a negative change in critical analysis is observed due to the dulling of the transient. Even after all the positive work with downward or upward compression and balancing the tone, there is only so much dynamic range reduction that can be achieved without looking at another way of shaping the envelope. Soft clippers add distortion to the upper transient, cutting them back while affecting the transient shape with distortion. This adds the treble lost through the dulling of the transient in its reduction if applied correctly.

Care must be taken however to avoid making the audio distorted overall, or too bright which is opposite to a limit that just dulls the attack of the audio if overly applied. Equally, a soft clipper with too much application will add clear distortion, especially audible in the Side element to start with, then move audibly to the whole sonic outcome. In beginning to comprehend how much is too much, it is important to first observe the link between input and output.

The link between input and output

When setting up a master, it has been previously discussed to create a source track as the start to the transfer path and a copy of this as the original source which will remain untouched throughout the mastering process. At any point, a comparison can be made back to the original. For this to be valid it is crucial to work at equal loudness. The source mix must be equal in level to our reference material. This must all be levelled equally to the same calibration being used to observe the numeric of loudness difference between these two aspects. The difference in perceived loudness in dB is between the reference material on average and the source mix. Depending on the mix dynamic range, this could normally be anywhere between −12 and −3dB as a theme. But it clearly depends on the dynamic range of the reference source used and the mix's own dynamic range. Remember, this is not about measuring output to a peak level, but levelling at equal loudness to hear the true comparative between mix and references while also maintaining correct listening level at our calibrated optimum. Inherently what is left is amplitude difference in peak between the reference material and the mix as the RMS is levelled for equal loudness. This makes for a true appraisal of the difference between the audio.

It is critical from the outset that this levelling is correct, because the difference is how much reduction in dynamic range will potentially be required to meet the same levelling at peak/RMS to our reference material. In approaching the gain structure of a setup like this, you are likely to avoid the temptation to push the output level too far when the mix sounds balance against the reference masters. Otherwise why push the master any harder if it sounds concurrent with other music and still sounds better than the mix at this equal loudness. As processing is applied, the output of the transfer path should maintain equal loudness to the original source after every process applied. In comparison back to the original mix, a realistic loudness balance will be achieved to make sure the audio is actually being improved and it is not just the effect of volume difference being heard. This will also positively affect any level normalisation scenario as the master will sound better than the mix and it will be more linear in dynamic and tone if these have been controlled effectively. It will be sounding better than the original mix at equal loudness.

This can be easily checked once the master is printed by level normalising offline both the mix and master created, readjusting both outcomes by the same amount to make the original mix volume back to the same level as it was in the original track path levelling. When comparing back to the original, both are relatively levelled to our calibrated listening level.

An example is levelling the mix to our calibration level on our RMS meter at say −18dBFS. Adding the references, they needed turning down −8dB from their original level peaking close to 0dBFS to match the same level on the meters RMS. Thus the SPL in the room is playing the music at our calibrated optimum SPL. This levelling should now be honed with critical listening to achieve effective equal loudness turning the mix up or down relative until equal loudness is achieved by ear. In doing so the mix was now half a dB up from where it was visually levelled. Remember, meters are a guide and our listening delivers true loudness comparison. The power of the mix and the references now sound equal and they are averaging around our RMS calibration at −18dBFS. The peaks of the mix are all over the place. The peaks of the references are uniform at −8dBFS because that is the value of reduction to our listening level. Completing the master while maintaining equal loudness throughout means the RMS average will have been maintained at −18dBFS and the mix would now have a reduced peak to RMS ratio. In playback the master will have a peak at −8dBFS. For this to be true the output of the transfer path must also be restricted by limiting at −8dBFS at peak, but sound is concurrent with the references in level as they are peaking at −8dBFS. The master should sound like it sits effectively with the references. If not, you have a tonal issue or you are not at equal loudness.

In level normalising the master and the source mix, using whatever standard, the mix will adjust in level as will the master. Adjusting the mix back to its original levelling at −18dBFS RMS and equally changing the master by the same amplitude as the mix, the loudness difference between the mix and master in the normalisation standard has now been established back to our calibration. This will have a dB value difference, but that is not really important. The comparative knowledge is found by soloing the mix against the master in this new gain structure. The master should sound better than the mix. If not, you are either compressing the master too hard or have a tone balance issue which should have been notable in a peak levelling to your references. Whatever the scenario, the master should always sound better than the mix.

This is a true test of whether the master has improved relative to the mix in density and tone. If your master is quieter, it will not sound better than the mix, even though the mix is imbalanced. It does mean that the master you have created is either too compressed/limited relative to the mix or the tonal balance is over focused in areas. This simple comparison is what will happen on ingestion to YouTube or any other level normalised system. For example, the band made a pre-release video using the mix or mix limited master from the mix engineer and your master is used for the official video. If this normalisation test is not true, your master will not sound as good as the mix version. In what world is that helpful to anyone?

Tonal balance and loudness

The correct weighting of frequency across the spectrum is critical in our ability to achieve a positively 'loud' outcome in both level normalisation and at peak 0dBFS output. Too much frequency distributed in the range below 400Hz will reduce the potential positive loudness outcome of the master. If ill weighted across the spectrum, the dynamic range reductions can only do so much before this imbalance comes to the fore. It is crucial to achieve a positive frequency balance from the outset. Having too much low frequency means the master lacks treble. Remember the yin-yang of tonal balance – one ill weighted action affects the opposite.

Ultimately, audio referencing is your guide to correct tonal balance. The more you can build your internal 'sonic library of listening', the easier it will be to comprehend what an appropriate tonal balance is. This is important for each piece of music in its own genre and their place in the wider musical context of all outcomes that music exists in day to day. The better your understanding of this average, the more effective your tonal balancing will be. Thus achieving apparent loudness in all regards with ease.

Start, end or both

To achieve positive peak reduction, the first aspect to understand is to not try and achieve it all in one process. This is bound to fail unless only a couple dB of reduction in processing is required. As a theme a couple of dB of reduction in one limiting process is enough because if more is applied, the negative effects of the limiting process will clearly be heard. Staging this dynamic reduction especially in the peak is key. There is no point putting one limiter after another as they are just targeting the same peak shape. It is best to achieve successive manipulations of the envelope shape. Applying a limit at the start of the transfer path, the peak to RMS ratio can be reduced a little. It is important to start to observe how this will work in conjunction with the following dynamic process which should not be a repeat of the original process. In this case applying another limiter is the same as adding more on the original limiter. In skimming off a few peaks in the song by a couple of dB maximum, skim not every strong transient but the loudest. The audio will now pass the threshold of a following compressor by a controlled maximum value set by the limit. This is more controlled in its range and potential response in the ratio of reduction. This should give a higher level of control over ratio, and more effective knee range can be set as the dynamic range is confined in the peak. This thought process could also be started with a soft clipper, the same principle applies; or a limit followed by the soft clip to gain more control on the clipper as it is restricted. Personally, I am not a fan of a soft clipper early in the chain because I feel they are more effective when the dynamic range is more restricted overall. There is less variance in the tone when the momentary elements are more controlled. It means the distortion created is more linear across the track so it is less noticeable relative to the amount of reduction applied. But there is not a correct answer. Just comprehending what could be done, testing and experimenting will help you find what you feel sounds best. What does not change is the principle of reshaping the envelope at each stage of reduction.

Either way, having controlled the initial dynamic peak, the following compression process will change the overall aesthetic, or an EQ process would adjust the density, meaning the envelope shape has changed. Maybe apply another post limit to this if the EQ has excited the transient as the shape has changed. Whenever using a soft clip to shape the peak, you can apply a post limit process to reshape this outcome, often a pre limit level to this would smooth the soft clipper. All these processes are changing the shape of the peak after every process, not repeating the same thing. Between all these small gains in dynamic restriction, it probably has achieved between −4 and −10dB of peak to RMS reduction. For example, the transfer path may consist of: Filters > Pre limit > Upward compression > Downward compression > Limit > Soft clipper > LPF > TP limit. Each stage achieves a little more dynamic restriction with little effect on the quality of the audio other than making it sound smoother, more glued and unmasked in outcome. It has been effectively mastered and is more compressed.

Obviously, this chain should be devised from analysis and not a 'one size fits all' transfer path. If that worked, computers would have taken all our jobs decades ago. Artificial intelligence (AI) uses this pathing premise, but it will fail to make the most effective outcome the majority of the time because it is not 'listening' to the audio. Though on occasion a path might just fit if the inputted 'listening' parameters (user defined genre, style, level, etc.) are close to being correct. But one person's perfect outcome is another person's far from correct. This is part of the issue with discussion around AI. A lot of the chatter is made by people who do not know what the outcome should be anyway. They can just hear something that sounds better, most probably because of the loudness difference. But this obviously does not mean it sounds correct in any sense or is as good as it could be.

Integrating soft clipping

This is now a mainstay of a modern sound. The nature of soft clipping is that it hacks off the peak information in the same way pushing hard into an analogue tool would. In doing so, distortion and harmonics are added, often this is also described as saturation. Depending on the aggressiveness of this push it equates to the amount of artefact created. Generally, the smoother the peak information going into the soft clipper, the easier it is to manage the correct amount of push from the clip. As a theme it is best to apply after a limiting process. This does not have to be directly before in the chain. It is just helpful if excessive peaks are controlled before the soft clipping. Because harmonics are generated by the clipping, a low pass filter (LPF) is often a good idea post the clipper, preferably high in the audio spectrum assuming the audio path is at the higher sample rates, perhaps at 28–34kHz on the filter to avoid tainting the treble but taming excess. This hems in any overly zealous harmonic content generated. The TC Electronic Brickwall Limiter 2 on the System 6000 has always been a very useful soft clip, however the newer TC electronic Brickwall HD plugin has a wider variety of settings opening up opportunities to use it more than once in the chain with less aggression. The Sonnox limiters have similar but more board band clip functionality. These tend to focus more of a change in tone on the mids and bass meaning sometimes these different processors can be used together.

Clipping the AD

This is something I have never had any positive result with in all my time using analogue paths and testing many converters. I do think that some engineers who discuss using this method do not realise their ADC has a soft clip/limit function, hence they are not clipping the ADC. They just like the sound of their converter's soft clip function. Digital clipping to me is not a healthy outcome in mastering. In a mix, on a single sound as a creative tool, interesting sonic results could be achieved. But I see little point in mastering. Though as always, it is best to be open minded and willing to explore a possibility of process again if a differing example comes to light. There is not a correct answer, just one that can be justified sonically in all regards.

True peak limiting

True Peak (TP) means the limiter is using the same principles as upsampled metering to decide on the amount of gain reduction required. Remember, upsampling enables a tool to see the potential positions of the audio between the actual samples. In upsampling the audio the potential peak overshoot which would be clipped at 0dBFS can be redrawn accurately because there are more divisions (inter-samples) to redraw the potential outcome. This is especially useful in determining if a lossless file pushed towards 0dBFS would clip if converted to analogue or a lossy format where the peak will not always come out in the same place as the lossless due to the encoding. If it did, it would be lossless! Hence if we use a TP limiter correctly, it can prevent clipping at the output of the DAC and stop some of the potential clipped outcome in the lossy file when processing from lossless. Obviously a safety limit for being more rigorous about the potential overshoot will apply a higher amount of limiting than a normal dBFS limiter.

All TP limiters and meters are not the same, testing outcomes will help you navigate but I have found the T2 limiter on the Weiss range to be quite effective. The TC Brickwall HD if not pushed to hard can work as a main limit as well. As discussed previously, staging out our dynamic processes would mean a TP limiter at the end of the transfer path will only be marginally reactive to the audio, at most 0.5dB at only a few points in the song. When setting your final limit process to capture those last peaks I have found it more effective to use a normal limiter followed by a TP to correct any overshoot rather than just using the TP. It sounds more transparent.

0dBFS?

It is critical to be mindful of the final output level. Setting the main outcome to 0dBFS means any potential clip down the line is significantly more likely than if set below this maximum threshold. −0.3dBFS was common practice in the heydays of CD Audio, to avoid any potential clipping on the output of poorer CD player DAC. But in a modern context, this is down to your interpretation of the outcome and requirements of loudness. -1dBFS would be suggested for some lossy conversion outcomes, though the industry is transitioning out of the era of lossy. A master is not just for now, but the future outcomes in audio. The trouble with turning our peak level down is the music is now quieter at peak loudness. This is just another aspect to consider in the delivery of a final master. This adjustment would have no effect on LUFS normalisation as long as the audio is turned down under normalisation.

In terms of the effect of peak on consumer playback, all DAC were not made equal. One converter's idea of zero is not the same as another's. The less expensive the DAC, the less accurate the outcomes. This was much of the rationale for −0.3dBFS peak level when mastering in the era of CD only before the principles of upsampling and TP metering and modern brickwall limiting were available.

A true peak limiter means you can push this last 0.1dB to the maximum potential. Though with a pushed master, an output level of −0.15dBFS should be observed to avoid a potential clip on the DAC output. This extra control of the final dB of audio is something not easily or effectively controlled with a conventional digital limiter process.

Having experimented with many limiters over the years, the standout tools for me when using TP are the Weiss T2 mode and TC electronic Brickwall Limiter 2 in upsample or TC electronic Brickwall HD.

Metering the loudness

I always have several meter outcomes observable at any given time, the most important being a final TP meter to gauge actual peak loudness on final print output. It is always equally healthy to observe the RMS of the audio to gain a concept of density relative to peak and how much the RMS (loudness) bounces in different sections. I think a LUFS measurement does this but not as musically. The measure I find most helpful to observe is RMS. These are just visual observations to help guide our listening perspective. The meter is there to help, not to make the decision about how our master should sound or how compressed it should be.

Summary

Key points to in navigating loudness:

- You cannot do it all with one limiter so stage your dynamics in the transfer path.
- Reshape the envelope before targeting the same area again.
- Skim off excessive peaks where you can.
- Use soft clipping after a limit process for a smoother outcome.
- True peak limiting should always be the last safety.
- Find your max peak output level, anything apart from 0dBFS.
- In evaluation observe the mix and master under level normalisation.
- During mastering observe the peak levelling relative to other reference material but do not 'chase' a peak loudness.

Lossy outcomes and adjustments

In adhering to the above, everything should navigate differing grades in quality effectively. If the output format dictates, it must be submitted at −1dBFS. I would still stick to the same premise laid out by just adjusting the final audio output to that level. It is also important to take into account that all formats are transitory, lossy will not be with us forever, hence the premise of −1dBFS in this type of delivery is going to disappear. HD streaming is now becoming the norm and lossy is falling by the wayside on many platforms. I would say it is now incumbent on a provider to prove their future potential earnings and market share based partly on a lossless model of delivery.

Loudness ingestion for post-production

In submitting audio for ingestion into a post-production scenario, there is a likely peak requirement of −1 or −2dBFS TP. But equally the LUFS is likely to be very low relative to the peak, −23dB LUFS. The peak is irrelevant as our audio will be more compressed than this for all good rationale in relation to music delivery. I think it is worth noting at this point in terms of resolution, finishing a master using all the bit depth in a file, whether our peak is 0dBFS or −1. Each bit is 6dB, so our 24-bit outcome uses all the potential dynamic range. If levelling down to −23dB LUFS, our peak may end up at −17dB with the weight averaging at −23dB LUFS. We have lost 2 bits in resolution or 3 if our peak was −1, meaning our master is really at 22 or 21 bits in a 24-bit carrier. This creates a good rationale to observe 32-bit delivery for ingestion to avoid any loss in the original quality before the final reductions for format are made in post-production balancing. But this is also down to what the company dealing with the ingestion will accept.

Why not LUFS?

Put simply, it is for post-production and multiple media formats. It integrates music with bass and no gaps, alongside spoken word with no bass and gaps. It works very well for this, but it is not a helpful measure with music audio only. I have always found a straightforward RMS meter gives the most accurate impression of musical loudness when working, recording, mixing or mastering music. But in the end, it does not matter what you use as long as you are listening and not just looking! Metering is always only a guide, and should be used as such. The observation from your listening is what really matters.

Ignore all of the above!

In testing and observing many commercial outcomes over time, there is little adherence to whatever assumed standard practice there is. Whatever the medium or fixed perceived standards in level, there are none that can be enforced on music delivery. There are plenty of successful tunes that are just clipping away at TP at 0dBFS as you will find in testing commercial outcomes. But just because it can be measured, does not mean it should be changed. I would just trust your critical listening and test, test, test to know what outcomes sound excellent to you on multiple systems. It is how the music sounds to the consumer that matters, and not how many different ways it can be measured.

Bring it all together
Engaging in mastering

This chapter is designed to help you focus and develop your approach to mastering, enabling you to deliver positive outcomes for you and your clients' audio time after time. To support this discussion, it is helpful to have a copy of the mastering analysis worksheet available in the online open share resource folder that supports this publication. See the introductory chapter for the link.

The aim of this worksheet is to help you construct an effective methodology to achieve a mastered sound. In time, with enough self-discipline and evaluation you will be able to enact the desired outcomes without the need for any supportive documentation. But it is initially important to consider differing aspects in approach and not rush to decision making. Without a full comprehension of potential methodologies, it is easy to miss taking important observed contexts on board. The purpose of the worksheet is to help you develop these skills with the knowledge and depth required. Developing critical analysis and evaluation skills is not restricting creativity in approach. Being dimensional in approach and development of methodology comes from in-depth evaluation and reflection around your own practice. A one size fits all will never wash in any part of audio production. The notion of an ideal mastering chain is equally nonsensical; every piece of music is different and requires a dedicated approach to achieve the best possible outcome. Obviously, there will always be themes where a similar approach will come up, but just because we can paint a fence doesn't mean it should always be the same colour.

DOI: 10.4324/9781003329251-16

The first listen

Our main consideration when starting any mastering project is the need to listen to the mix master at an appropriate level. As discussed earlier this is around 83dBC SPL in the sweet spot. Though if you have digested the knowledge to be gained and conducted the experiments suggested in the initial chapters, you will have found your ideal listening level. Using calibration on your monitor system makes finding this level quicker to achieve. Being conscious of loudness calibration and dynamic range is crucial to achieve optimum equal loudness throughout the mastering processes. Remember, if you do not do this you will be fooled by volume difference, whether in the application of a tool or comparison to the original mix or audio references.

When starting the process of analysis, the source mix must be levelled at an appropriate loudness. Do this with the RMS meter using the visual scale only with no audible playback. It is just getting the mix in the ballpark for appropriate listening. Once levelled to the chosen calibration, I would listen and observe the whole piece of music from start to end in one pass, much as the consumer would when listening to a song for the first time. This initial observation can be the most important as it will have our gut response to the music, not in terms of how we might like or dislike it musically, but our first response to the sonics.

For example:

- Did the vocals sound appropriate in level, hold position in the depth of field?
- Were instruments masking anything or masked by something else?
- Was this in a specific frequency range or the whole instrument range?
- Was the macro dynamic between sections ok?
- Did it just lack clarity/transparency?

Remember these observations are about the sonics of the music within itself, not an observation of how it translates in tone or loudness next to music in general, that is where references come in. But at this point, you should be listening for how it sounds relative to itself. For example, does the bass end sound too full relative to the rest of the mix sound balance? Are aspects overly dynamic next to other instruments causing masking and so on?

It is also positive to note aspects positively viewed at this stage in our first listen.

For example:

- The mix has a positive stereo image.
- Bass and kick have clear focus.
- General transparency is good.
- Instruments are clearly presented in the sound balance.

In doing this, make a note in task 1 on the mastering analysis worksheet of all these gut observations. This can be referred back to at the end of the mastering session to make sure negatives are fixed and our processing has not disrupted any of the positives. All the aspects should sound as good or better than they did. Otherwise, our job of improving the outcome in all regards has not been achieved. This is the most important evaluation of a master relative to mix, it must have improved in all aspects, no ifs or buts.

At this stage do not make any link to processing in our notes. Observe how it sounds, not how it could be changed. For example, note 'The vocal dynamic is too large', do not write 'The vocal needs compressing'. It might well do but it is important not to jump ahead. Being able to see the full context of the analysis helps to make more effective tool selection.

Make all the notes around your observations in task 1 on the worksheet. To help your focus there are three headings to consider observations around: stereo and spatial impression, tonal balance/timbre, and sound balance and the dynamics in the micro and macro. It is not important to respond to all of them, just note what you have observed under the correct heading.

Comparative analysis: Loading the reference

The next practical outcome is to load your chosen audio references to another track in the DAW. Avoid overlapping with the mix master in playback. Load sequentially, one after another with a gap between all files. Level these mixes to the RMS meter calibration setting you are using. This can be done with the monitor controller muted again to get the audio near to the correct listening level before levelling by ear. As they are mastered and appropriately restricted in dynamic range, levelling of their RMS to the meter should be straightforward in relation to the mix where the dynamic range will be less controlled. This reference levelling is more accurate to your target loudness in SPL. Now level the mix to the reference by ear; it is the mix that should be adjusted in level to achieve the appropriate levelling and not the reference. This is because the references are more accurate in their RMS density and have been mastered, and will achieve a more effective levelling to the calibrated level.

When considering the inter-song levelling, it is best to focus on the common element, primarily this will be the vocal or solo aspect. The vocals should sit at the same point in the depth of field, one mix vocal should not jump forwards or back in the spatial impression when A/B'ing between the two. Remember, amplitude controls the relative level of a sound to another sound in a mix, hence the louder sound will sit in front of the quieter. The same principle is at play here; it is just auditorily more complex and the music is different. If you focus on the vocals, you should clearly hear when A/B'ing they do not move in that depth of field perspective between each other when at equal loudness. That way it is clear they are the same loudness relative. Do not be distracted by the power of the tone. Once levelled, it is the true tonal difference being perceived. The meter is only a guide to level, to level between music it is critical to use our ears to decide. Meters are just not accurate enough even within the same genre. Also do not be distracted by the mix being more dynamic, you must take an average of its relative level to the more compressed audio reference. As said this is not straightforward and the sources are complex in their sonics, but relaxing focus and listening in overview you will start to clearly hear these aspects at play.

This comparative between mix and reference is best achieved when running the DAW in playback and flicking the playback cursor between the audio in similar sections to achieve the comparative. As discussed with step controls, changing from one fixed value to the next fixed value gives a true A/B. The same principle is at play here. In wanting to compare between the reference and the mix directly, it must be conducted in similar sections of loudness, i.e., there is no point comparing the mix intro with the audio references main chorus. Once levelled correctly, the true tonal and dynamic difference to the reference will come to the fore as there is no longer a loudness difference. It is common when conducting this task to notice the audio changes in tone in our perception as the loudness difference changes. This is normal, it means the volume is changing our perception of comparative tone in looking for that perfect levelling. I cannot overstate how important this aspect is. If rushed, getting it wrong, all following observations will be influenced by volume difference and hence any contextualised process will be incorrect. You'll be fooled by the loudness difference and not the actual sound in the difference between the music.

Now in task 2, make the same observations about the characteristic of the mix but based on a comparison with the references. These are essentially correct in stereo/spatial impression, dynamics and sound balance in both micro and macro, and lastly the tonal difference in overview. I think the tone is best viewed as imagining the reference as flat across the spectrum (a straight horizontal line) to visualise how the mix appears in difference. Is it dull in the treble relative, inflated in the bass, overly bright in the mids? Consider what shape these observations are to build up a picture of the mix's tonal and dynamic difference across the spectrum in overview. Remember to think about whether the observation in tonal change is momentary (dynamic) or static (fixed) in amplitude. Sometimes it can be helpful to draw this out to visualise as in the two examples in figure 15.1, where the top analysis shows a mix lacks bass and treble statically which emphasises the mids relative to the flat reference, and the bottom analysis shows a mix that has momentary difference in the bass and treble but lacks mids overall.

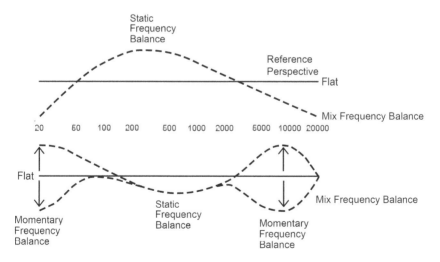

Figure 15.1 Distribution of tonal differences from a 'flat' reference to analysis of two differing mixes

Constructive audio analysis: LRMS

Once all our observations are noted on the analysis worksheet, move on to task 3. Use the monitor controller to consider the matrixed audio paths to ascertain where elements of interest are placed in the mix, i.e., Vocals, Kick, Bass in the 'M', Guitars, Synth, ambience in the 'S' and so on. If there is no obvious difference in L and R, make note of this too. It is not unusual to have a tonal imbalance in guitars or the metalware of a drum kit bias on one channel rather than the other of the mix. Quickly flicking between L and R during playback in mono maintaining equal loudness in comparison across both speakers can make this difference obvious. The point is to make a clear breakdown of which path a given aspect of the audio mix is mainly contained within when assessing how the mix has been constructed in its stereo and spatial impression. This analysis could be used to better target processing later on.

It is also helpful to make an observation of the Mid/Side to enhance our tonal observation and visualisation in task 2 to highlight how the difference may vary across the stereo impression. For example, the mix might sound as bright as the reference but lacks width in the stereo impression in comparison. Listening to the difference it is clear the Side is dull relative to the mono of the mix and our observation of the reference in the Side. The lack of width could be tonal and not just amplitude.

Corrective analysis and analysis in overview

In task 4, take our observations noted in tasks 1 and 2 and think about the context of each observation. With each note see if a specific element in the mix observed needs correcting, or was it something that could be defined broadly as an overview of the whole mix. For example, a corrective observation would be that the vocal is sibilant. An observation in the overview would be that the mix is dull. It is crucial to separate our examinations at this stage to define what observations require 'corrective' processing. Something that needs fixing to get the mix in the right place to then applying the processing needed that was observed in 'overview'. Often a good way to do this is to note whether the observation was about an individual instrument (probably corrective), or the whole section of the music (probably overview). You need to order your observation into these two categories. Do not worry if everything is just in one, as long as you have considered each observation in turn. For example, if observing the kick had too much power, this is an observation about a specific instrument in the mix that would fall into the corrective processing category. In contrast a mix having too much low end would be listed in the processing in tonal and dynamic overview.

Contextual processing analysis

Moving on to task 5 on the worksheet, the focus is now on the need to link a process that could positively affect each observation made.

If the observation was 'Vocal is overly sibilant', this could be corrected with a split band compressor or what would more commonly be called a de-esser in mixing. Also a multiband could be used with HS shape or bell to target the focus of the sibilance. Because this observation was about a specific instrument, it would be assumed this was placed in the corrective processing category in the previous task. As another example, 'Dull in tone' could be linked to the simple application of a HS boost on an EQ. It would be helpful to consider an LPF to hem this in as discussed in the chapters about equalisation.

Other examples could be, 'Lacks excitements in top end/sound masked/ dull in the tops', which could be linked to a HS but with a dynamic element, change the top end on the momentary aspects to excite. It could be dynamic EQ, but could be equally achieved with a multiband expander. The choice would be in the shape of the filter required; this knowledge comes back to testing tools to comprehend the difference in slope of filters on differing processors to understand what is required.

At this stage it is best not to get too in-depth noting a specific tool, like a TC System 6000 de-esser, or EMI TG12345 HS. This contextualisation needs to be in the broadest context of a tool directly linked to changing the observation made. For example, if the observation was:

'Macro dynamic to large in the intro/first verse' > (link would be to) Manual compression.
'Overly dynamic in the micro throughout' > (link would be to) Downward compression.
'Masking in the quieter details during breakdowns' > (link would be to) Upward compression.
'Low mids have too much focus' > (link would be to) LS+ and HS+ and HPF/ LPF to hem in.

Each time we are looking for a simple response to each observation.

Transfer path construction: Contextualising constructive analysis

Moving on to task 6, now the focus is on the need to link our observations in task 3 around LRMS to the individual processes noted in task 5. If an aspect being targeted is in a discrete path of LRMS, it should be noted for processing in this path, contextualising the analysis from constructive analysis. For example:

'Vocal is overly sibilant' > Split band compressor HS.

If the vocal observation in our LRMS analysis was in the mono element, it should only be processed on that path.

'Vocal is overly sibilant' > Split band compressor HS > M

Contextually it would also be interesting to think about this relative to any backing vocals. The sibilance is in the mono on the main vocal, but if some of the backing vocals are panned in the 'S' are also sibilant, it could be possible to consider:

'Vocal is overly sibilant' > Split band compressor HS > M and S unlinked SC.

If the sibilance in the S was in sync with the main vocal, it might be worth considering linking the SC to achieve a more effective trigger, as the main vocal sibilance is likely to be stronger. Using the M as the main trigger for the S as discussed in Chapter 11 'Practical approach to dynamic controls' in section ix. 'Sidechain compression'.

'Vocal is overly sibilant' > Split band compressor HS > M linked SC.

Another example of the observation was:

'High hat too bright' > Bell EQ.

If in our LRMS analysis the hats were prominent in the right, the processing could target the channel two path.

'High hat too bright' > Bell EQ > R

The control on the brightness in this path will make the overall tone better balanced, especially in the imagining of L and R, rather than if processed the stereo where the imbalance would still exist.

Do not forget that the stereo path is there, as well as an outcome. Do not get carried away with discrete processing. There is nothing wrong with stereo if that is where it has been best observed.

'Overly dynamic throughout > downward compression > stereo SC linked.

Contextualisation of all our dynamics SC should be considered in this task. Should the process be triggered by the same routing as the original signal?

Guitar's masking drums > downward compression > Stereo > SC M.

The trigger of the drum peaks will dip the compressor into the guitars in the S. Thinking back to our discussion about the SC running into the detection circuit of the dynamic, it is also helpful to note the potential crest setting. In the latter example, the trigger would be from the peak, the drum transients. But if targeting a vocal in the M, there would be an HPF LPF around the voice on the sidechain and crest set to RMS to avoid any percussive sounds. Also remember an SC is an audio bus, hence considering if a pass filtering is needed on the SC to achieve a more effective trigger with compression avoiding the low-end energy can be effective. Once establishing the routing channel path for the audio and the SC, move on to task 7 and consider the tool type.

Linking tool type to processing

Looking at the whole aesthetic of the process and its purpose, consider each outcome to attach a tool type to the processing. If the intention of the process is not to be observable, correcting something in the mix, this would direct us to a clean tool type or even linear phase. If the expected process is adding value such as an EQ boost, a tool type that adds saturation/harmonic content would be effective, this could be as far as a valve EQ, but an analogue or analogue modelled EQ with positive phase distortion would add interest and focus to the addition meaning less of the boost needs to be applied. The consideration is whether the tool should do its job without artefact or add some of its unique colour. At this stage, I still would not make a specific link to an actual tool or plugin but rather think about the aesthetic outcome required. For example, if the observation was that the mix was dull, our thought process would be: Dull overall > HS boost > Stereo > Valve. This should not be linked to a specific tool like a Manley Massive Passive or a Thermionic Culture Ltd 'The Kite' directly.

Throughout this analysis method and contextualisation of tools, focus has to be on sonic principles and theory. The 'why' and the 'how', not the gear or the studio being used. I think it is a very important aspect in developing your comprehension of tools and processing and equally making the skills developed transferable, by not becoming reliant on a piece of equipment to construct a path, but the knowledge around how you want to change the sonics to achieve our outcome. You can respond to the observations made and not be reliant on a fixed set of processes or pathing predetermining decisions.

It is not about the gear, but the knowledge of how each process can change the sound. In doing this you will also make more interesting connections in how outcomes could be achieved. For example, in our evaluation, a valve compressor would give the glue required, but this is just stereo compression with the addition of saturation and harmonics. As an analogue tool, this may have restrictions on the pathing or how the SC can be accessed, which also might have been part of how and what was observed could be changed about the sound. The decision is a compromise in the pathing and SC, or think about how else to achieve saturation and harmonics. For example, a soft clipper following or fronting a clean digital compressor process achieves all of our aims. Both of these aspects would work, but the knowledge to consider them means a better selection will be made.

Evaluation of the transfer path

Coming to our final task 8 and taking all the outcomes and consideration up to this point formulated in task 7, we consider how all the processing and tool types affect each other. What is the consideration of processing order? Are there any duplicate processes? Is too much additional colour being added to the overall aesthetic from our tool types or is there not enough? Are three valve processes really necessary? Does one just not add what is needed? Have the extremities for the frequency range been pushed too far, say a low shelf pushing the sub, or is there a pass filter to hem this in? It is time to start thinking about the overall pathing and what effects each process has on the following processes?

In part this started in task 4 when ordering our analysis into corrective or the observation in overview. Assuming the corrective 'fixing the mix' aspects would be first, but equally normally, a pass filter would be at the start of a chain as it is removing energy in correcting. But if the pass filter is there to capture the harmonics created by soft clipping or a valve processor, it would be required post to these processes. Dynamically, ordering needs consideration, the more restricted the dynamic range becomes, the smoother the threshold range. But equally this makes it harder to achieve a trigger on individual aspects. In the same way if a boost is made to the EQ before the compressor, the change is pushed into the dynamic restriction. If applied post, the sound is more even and the EQ will sound smoother by consequence. Both sound different, the choice is an aesthetic one, not right or wrong. The important thing is it is explored in principle rather than just plugging in gear to see what happens!

There is not a correct answer but there is one that can be justified. If you cannot explain what a process is there for, it should not be in the chain. If you cannot justify why the type of processor is being used, it should not be in the chain. You need to be honest with yourself and really justify each and every selection. This is why conducting analysis and contextualisation to potential processing is so valuable in starting to appropriately link all the parts of the chain together. Make the processing path link directly back to our analysis. Create an audio outcome that sounds better in all regards. This is mastering.

Once formulating a path for processing, it is worth noting that up to this point the only actions taken have been to listen to the music of both the mix and reference and consider our analysis. At no point has our favourite EQ been inserted and swept through the frequencies playing around with the frequency spectrum, or a compressor or limit applied to the source mix to match its level in loudness to our reference to 'listen' to how it sounds. All this is nonsense and poor engineering practice. If a piece of equipment is to be understood, it should be tested. Its sonic outcomes have then been acquired to become part of our knowledge. This should not be done while mastering. Our focus needs to be on our unique piece of music and keeping our listening focus on making observations and decisions. In conducting this analysis, contact with the audio has been minimised thus avoiding becoming overly familiar with the mix by over listening. Our fresh listening perspective is now a clear and ready to consider approach to application.

Application

Each part of our formulated path should be clear in our understanding of what the process is, with what and how it is routed, and most importantly what it is meant to impart on the sonics. In application, each process can be applied in turn and its effectiveness evaluated. It is okay at this stage to tweak, but if something clearly does not function in the way it was assessed, it should not be applied, and it can be considered again later in evaluation. At this stage, mastering is often a lot of small changes that add up to the overall aesthetic difference which is normally now quite a considerable change. Do not be put off if the focused difference in the application of a single tool only makes a tiny difference. It all adds up!

Equally part of this approach is about improving our listening skills, the more observations that are made around these small changes with focus towards what should happen and why, the more these changes become obvious. You just need to stick at it, stay on the processing considered and apply those outcomes. Once the path is completed, it is then possible to conduct the final and the most important part of any mastering.

Equal loudness evaluation

In this final reflection, look back at our notes from tasks 1 and 2. These observations were made in context of the music within itself and compared to a reference to give a full perspective around our perceptions of the mix. These positive and negative notes should be improved in all regards when listening to the master versus the mix at equal loudness. If the audio in all aspects has not been improved, there is a need to go back to the transfer path and augment to correct. It is critical to make good on our observations or we are not being honest with ourselves. There is no truer comparative than equal loudness. If the audio is improved in all aspects, make more comparatives, as suggested in Chapter 14 'Navigating loudness', regarding other potential levelling outcomes. Improvement must be achieved to be honest about the application of processing. If you have not improved in some regard, keep turning off aspects of the path that relate until you do work out where the audio is being inhibited. Look back at the worksheet and start again from there. The more you validate your outcomes, the less correction that will be required and the better the outcome that you will create overall.

The reality of mastering

If you have followed and completed the steps above, been critical and reflective and most importantly honest in your appraisal and approach, you will have improved the audio. It will sound sonically more balanced and together, but it is unlikely to be finished. The reality is without a highly attuned ear, it is very difficult to hear through the mix and address all the aspects in one assessment. In developing your skills, I feel it is best not to try to jump ahead and apply aspects without a direct link to analysis. Once you have completed the assessment and evaluation, take this output and regard it as the mix master_02 source and once again go through all the stages of assessment and contextualisation of your new analysis to tools. When the music is more controlled, you are likely to notice different aspects in comparison to the reference you can now improve. These are now much more likely to be in the 'processing in overview' category discussed in task 4. The obvious mix issues (corrective processing) will have been fixed in your first pass. It is where the aspects of tonal shaping with EQ and increasing density with more targeted dynamic reduction such as soft clippers and/or limiting will come into play.

If you are being diligent in adherence to your analysis, this new path constructed and processed will again improve the context of the sonics. What you will now be able to contextualise is both transfer paths from both rounds of analysis and process it together. Now looking at it as one whole path, are there any duplicates or ordering of aspects that could be changed/switched-out to improve the cohesion of your processing? In doing this, you enable yourself to see the bigger picture. Comprehend in better detail where the issues lie and how to effectively improve with processing.

Do not now restrict yourself and assume you are complete. Take this outcome as the source mix master_03 and conduct analysis again, contextualise to processing, apply and evaluate then reflect on all the audio path as a whole. Continue with this approach until the mix now sounds mastered and sits next to your reference material in context. Maybe it won't sound as sonically superlative, but it should not sound out of place. Your references are more than likely to have been superlative in every stage of the production process which is in part why they sound great! Remember, mastering can improve an outcome but it cannot change it into something it never was in the first place. But there is an expectation of a master to sit next to other music without the user having to reach for the volume or tone control to be able to listen to the outcome. You are looking to make the mix sit in this mastered context.

Restrictions of delivery formats

Convenience over quality: A century of consumer audio

As mastering engineers, we pour our heart and soul into the quality of the signal path, constantly testing and developing how to maintain the best outcomes throughout any engagement with processing or outputting. But it is critical to consider how others who are not audio engineers might want to consume our audio output. That has nothing in general to do with our perspective. Consumers, quite rightly, are just interested in listening to music. They are not on the whole considering aspects like quality and fidelity of their playback system, factors that audio engineers would see as important. Quality is a factor, but usually lesser on the priority list to cost, and ease of use and kudos. Historically this is also true, as a theme, the majority of consumers have always chosen the option that is easier, the most straightforward or cheaper.

Looking back over the history of delivery formats for music, there are a litany of duffs, not because of anything necessarily negative about the format, but it is inconvenience relative to other formats, cost and usability. For the majority, the tools that everyone would engage with were best for them.

Lossless

Delivery in lossless has two fundamentals to observe as Linear Pulse Code Modulation (LPCM) digital files resolution is defined by bit depth and sample rate. The facts about comprehending resolution change can easily be

DOI: 10.4324/9781003329251-17

learnt by testing. All samplerate convertors do not work the same way and thus sound different. Aside from this, they also have differing outcomes on headroom at peak and TP, so you need to convert, test and evaluate. In doing so, build a picture of what works well and is the most effective for your outcomes.

Truncation of bit depth on the other hand is nasty, it introduces harmonic distortion. If a 32-bit float file is converted to 16-bit, truncation takes place. Same to 24-bit; anytime the bit depth is reduced, truncation takes place, i.e., hacks off the rest of the previously available dynamic range. There is no option to this action because by removing the bits there is nowhere left to store this data. To mitigate this effect, dither needs to be applied, a low-level noise to randomise the last bit, and in effect trick a higher variation in depth into the smaller numeric value. There are many different dithers, they all impart a different sound on the audio. I have dithers I prefer depending on the application. This knowledge is gained through testing, and you will gain the appreciation required to start to comprehend formatting outcomes.

Noise shaping dither tends to be best for compressed music as these push the noise out of the audible spectrum. But more dynamic material, folk or classical often benefits from a broadband noise that can be perceived but overall has less effect on the tone. Personally, I find the POW-r dithers are effective, but equally keep on testing as different options become available.

If you want to take a master from a higher to lower resolution in LPCM format in the digital realm, you need to test the sample rate converters and dither options and be conscious of the headroom change at peak and TP. Also be aware of what happens to the tone, be sure to correctly moderate for equal loudness in comparison back to the original master in testing. Use sum/difference to achieve the best null in this regard.

Alternately analogue can be used to make this conversion utilising independent systems, one to playback and the other to capture. Personally, from experimentation and especially in the modern context of TP control, this is not my preferred option. Conversion from the highest resolution in the DAW to the lower gives a more controllable and repeatable outcome. Also take into account the need to often make multiple versions of a master, these are referred to as its variants.

Coming back to the premise of the restriction of formats, other than resolution change, there are no restrictions in digital lossless format other than that resolution dictates the available dynamic and frequency range available.

Lossy

Generally lossy formats are not being supplied as master file sets to ingest, this conversion happens on the lossy delivery platform. But references for clients or for promotional purposes are generally required. I would not tag lossy files with 'MASTER' to avoid any possible confusion, whereas 'PROMO' or 'NOT FOR PRODUCTION' makes clear what these files are. Obviously, it is important to be aware of the conversion of our high definition (HD) masters to lossy as most streaming platforms engage in this outcome, but this should have been part of the mastering process by engagement in correct metering practice observing peak and TP. There are plenty of other meters that will evidence conversion to lossy. Fraunhofer Pro-Codec or Nugen MasterCheck Pro do this job effectively. I am concerned with making music sound its best, and am not purely guided by what a meter might display. I think in this regard you should test and listen and not be worried by or be beholden to the meter or metering system. Lossy is used as a format, but is not ideal from an audio perspective, convenient for data rates/storage, but the future of audio is one where lossy will be less and less of a concern. But our masters will still be out there so there is a need to also consider future proofing our outcomes.

When having to convert to lossy, it is best reduced from the highest resolution format, as this will always give the most effective conversion to any lossy format avoiding dither or sample rate conversions from lesser LPCM versions. Ingest to some platforms such as Apple Music will accept an HD source for the primary delivery for both lossless and lossy streaming which is converted to Apple Audio Codec (AAC). Mastered for iTunes (MFiT) and its new incarnation Apple Digital Masters (ADM) is a formatting request to make the conversion from a higher resolution to AAC and the HD source having been checked by the mastering engineer to avoid any clips in the conversion process. This can be tested with varied tools, but Apple has an 'afclip' report which can scan a file to deliver a listing of possible clips/overs that happened in conversion. Some mastering engineers just set the master to -1dBFS TP which means it is almost impossible to register a clip on an afclip report without having to conduct a test or it can be tested using a real-time analysis. But this does change the context of the master playback in comparative loudness at 0dBFS playback, the LUFS normalisation will remain the same loudness. Again, it is another consideration and maybe compromise is required because of the restrictions of the delivery format.

For example, Apple will use the same source file for lossy and lossless distribution and streaming.

DSD/DXD

This format was originally designed for archiving audio, reducing the data required and achieving a more effective capture than Linear Pulse Code Modulation (LPCM) used in digital systems day to day. The first commercial consumer format use was Super Audio Compact Disc (SACD) which was meant to supersede CD Audio, but this was never fully embraced by the industry or consumers. That said, DSD is a growing market in HD download especially in the classical and jazz genres. This is because often in recording in these genres, the principle is to capture that performance in time and spatial environment. It is captured in that moment so someone who was not there could experience it as if they were part of the audience. This often means little or no processing needs to be applied other than a relative amplitude manipulation to achieve the desired outcome. This brings us to the problem and the positive with DSD – it is a one bit system, in a simple sense it means each sample stored can only have two states. These are: go up or go down. If the sample rate is fast enough 2.8224 MHz for SACD, way beyond anything in normal LPCM at 64 times the rate of CD Audio, each time a new sample is taken it just goes up or down relative to the next. In many ways it can be thought of as redrawing analogue. This has none of the negatives associated with a normal bit depth. But the super fast sample rate means the recording cannot be processed by a normal DSP, the most complicated DSD mixers only contain amplitude and routing manipulation, so a way of processing needed to be devised. This is Digital eXtreme Definition (DXD) which converts the DSD to 24- or 32-bit at the lower sample rate of 8 times CD Audio at 352.8kHz. This is also available as a consumer download format. You require a DSD/DXD converter to play these files but most high-end monitor DAC enable this mode. The practical inability to process the source recording has meant this still remains a niche market that justifies the outcome when practical use can be made of the captured source, i.e., there is little or no need to process what has been recorded. You could capture the final analogue mastering path for a metal band or a Korean popular music (K-pop) album in DSD, but the marketplace would probably dictate the final consumer outcomes as CD Audio, vinyl, cassette and digital in lossy and lossless PCM formats. The market for a DSD release would be very niche indeed. In the correct genres this is a productive and positive revenue generator. If you are interested in these areas I would suggest you investigate Merging Technologies Pyramix system which has been at the forefront of this technology and its development.

Vinyl

In contrast to digital, vinyl has many restrictions due to the nature of the physical format. Adjusting the master for these conditions imposed by the medium can change the tone and musical impression of the master, depending on the dynamic range and stereo image considerations. This means any change relative to the digital master outcome needs careful consideration on balance for all outcomes. These restrictions are focused in three areas: stereo width and its phase, macro dynamic variance and the harshness of the top end. I respond to each of these in turn below with practical approaches to address.

The first aspect that needs to be controlled to achieve an effective cut is moderating the energy of the bass in the stereo image. Really this needs to be mono below 100Hz for many cuts to avoid too much vertical motion with the cutting head. The lateral motion of the cutting head is mono 'M' and excess amplitude in the 'S', the vertical motion can cause break through or lift off during the cut. As the name suggests this means the cutting head goes outside of normal parameters and physically breaks through the lacquer, cutting too deep or lifting completely off the substrate (lacquer) during the cut. To avoid this, the low end can be controlled with an MS matrix and HPF on the S to create what is referred to as an elliptical EQ. This principle is explored in detail In Chapter 13 'Practical approach to equalisation' in vii. 'MS elliptical EQ'. Alongside this there are the obvious issues of phase coherence; out of phase elements will not cut. This means the correlation between the L and R must be maintained in a normal range. The use of phased elements to create pseudo width in a mix are not helpful at all and should be avoided.

The second aspect is the general width of a mix, causing issues in the vertical motion leading to break through or lift off. Either MS compression targeting the ill-weighted source on the S, or manual compression can be utilised to effectively control. These techniques are discussed in Chapter 8

'Dynamic processors and tonal balance' in section x. 'Manual compression in MS' and in Chapter 11 'Practical approach to dynamic controls' in section xi. 'Mid Side application principles'. It is also worth noting the excessive phase difference in any frequency range with the stereo image means the cut will fail. In the same way mono compatibility is important in a modern master, it is a fundamental requirement for the cut.

The third aspect is harsh treble such as vocal sibilance, brass, metalware of drums/percussion. Any sound that creates intense treble where there is a signal inside of the signal would be observable in the same way as in a frequency modulated (FM) signal generated waveform. If you have ever looked in detail at the waveform of these outcomes, you would see this intensity. It makes the cutting head distort because it cannot respond fast enough to the intensity with the amplitude. The same premise applies with clipped or heavily limited audio where the top end distorts, creating these intense waveform responses. With the normal excesses of sounds such as sibilance, there is a need for a high frequency limiter (HFL) to hold this area in for the cut, making the motion of the cutting head less erratic. In a modern context this would be referred to as a multiband HS. This principle is discussed practically in Chapter 8 'Dynamic processors and tonal balance' section vi. 'Multiband or split band compression/expansion' and Chapter 11 'Practical approach to dynamic controls' section x. 'Split band principles in application'.

To respond to the over-limiting or clipping, simply back off the level and ask yourself why you drove the level so hard in the first place! There is no need for heavy limiting to control a 'maximum' or chasing a level in digital, let alone for a vinyl cut. Vinyl's loudness in playback is dictated by the physically available medium which is reduced by the RMS density, the amount of bass and the running time, not how limited it is. This is counterproductive and only affects the cut negatively.

To me these three points of potential application are aspects that should have already been considered. The positives of digital formats are the lack of these restrictions of the medium, but this does not mean 'stereo bass' is good, or harsh brass is helpful for the listener and so on. I feel these historical restrictions of this medium, which has delivered millions of songs loved by billions of people, is part of our modern listening context and it is important to be aware of it in our consideration, especially if the artist/label intends to press vinyl as part of the release. This is not to say there is a 'correct' way to 'format' music, but holistically we must take in and respond to all these variables. Building your library of listening is key to knowing what will work and what will not. This is something that a cutting engineer is expert in; it is another whole skill set aside from making the audio sound great as a mastering engineer in the first place.

With vinyl cutting, you cannot miss the importance of the Recording Industry Association of America (RIAA) filter curves. Without this development, which was adopted towards the end of the 1950s, there would never have been the LP record. This pre-emphasis curve is a tilt in the audio spectrum shifting the bass down and the treble up by twenty dB in both directions. This obviously means the longer wavelengths are significantly less and the transients are enhanced in its aggressiveness, making the cutting head respond more effectively to the treble and the reduction in wavelength, significantly increasing the available time as the pitch is tighter. The pitch control during cutting controls the spacing between each groove, and if the relative amplitude is reduced, the pitch can be tighter and hence a longer cut can be achieved. Having cut the lacquer with an RIAA curve, the consumer in playback must use the opposite RIAA curve, a de-emphasis curve. This is built in on 'phono' inputs on Hi-Fi units or a DJ mixer. For a better quality sound, you can buy dedicated RIAA de-emphasis units or create your own curve. It is not unusual for the cutting engineer to tweak their RIAA to get the best out of their cutting head or the same for the transfer of LP sources in remastering.

In a modern supply context, there is no rationale not to supply the master for the cut at the highest resolution. A split 'AB' wav file format is the preferred outcome. The term 'AB' does not mean an A/B comparison of the audio but refers to the two sides of the vinyl product Side A and Side B, hence AB. With a two-disc long play (LP) record, this would be noted as ABCD, e.g., Side A, B, C & D. A complete sequenced single wav file for each side has pause transitions and fades embedded. This avoids any potential error in ordering or track spacing or even levelling errors. These AB splits should be supplied with a track list denoting the sequencing of each side A and B, or ABCD if it is a double album, assuming this would be HD wherever possible. The dynamic range of this master version is down to discussion with the cutting engineer and the artist/label requirement. Another master path print is another cost and equally another different version of the outcome. Whether this is a good thing or bad is subjective.

It could be suggested that consumers may like the vinyl versions because of the processing that needed to be applied for those restrictions of vinyl.

But they could equally be applied to the digital master, so both are the same sonically in delivery other than the physical errors in the production of final vinyl consumer format from cut to manufacture and playback. The same goes for the digital if the consumer is listening in lossy. The restriction of formats have their part to play in mastering, delivery and consumer playback. It is the mastering engineer's job to be intrinsically aware of all these potential outcomes.

Future paths with multi-channel audio

Multi-channel surround sound music formats for the general consumer have been a recurring theme in audio over many years, from ambisonics principles, Quad, encoding to UHJ, DVD Audio in 5.1 and now to Dolby Atmos with Apple Spatial Audio. The previous formats for music have never widely caught on for music only consumption. It maybe the same outcome for Atoms/Apple Spatial Audio for purely music listening delivery. A consumer just wants to listen to a song and feel its power emotionally from the melodies and lyrics and changing the listener's mood/perspective. Both of which do not require surround sound to appreciate, neither is stereo required. Many listeners use mono playback to listen to music though we do have two ears, and most appreciate sound coming from more than one source. But with more than two sources, inherent phase issues smear the audio unless set up correctly, which brings us back to two or one speaker as it sounds more coherent. Not forgetting headphone use, especially with the advent of better outcomes and usability in the last decades does mean many observe stereo or binaural encoding of surround formats this way. But encoding can lead to phase issues, smearing and image instability unless the product was recorded for binaural or is matrixed from an Ambisonics and not surround sound source. This is in part why these outcomes are not the norm, coming back to convenience over quality. If people do not feel an added value, where is the incentive for a consumer? Most just want the music to come out of a playback device. Our job as the mastering engineer is to make sure it sounds as good as it can when it does.

There is one format, stereo vinyl, that did easily catch on, because by design it was 100% compatible with mono vinyl. You can have a stereo deck and play mono vinyl, and have a mono deck and play a stereo record. With convenience over quality, listeners could play whatever and however they wanted. As a supplier, in time this leads to stopping manufacturing mono because stereo is mono and so on. Stereo is the source being supplied more and more by studios so the transition over time was to manufacturing stereo vinyl for the majority of releases.

In essence as a consumer, if a listener can hear the tune and lyric, job done. Who cares about the format. In the end the Queen and the King of music are mono.

Virtual reality (VR) utilising ambisonics and binaural playback configuration are very effective at linking sound to the visual aspects in view. Though music as an ingest into this soundscape is often most effective if the playback source has context in the environment, which leads us back to mono or stereo playback from a point of source. Or the music is just simply playing stereo in the headphone separate to the visual environment as a differing layer. At a virtual gig, the sound is coming directly from the stacks left and right of the stage. Unless we can walk between the performers where a control of the mix would be required, mastering would not be a consideration on the outcome of this type of production other than some master bus control, which would have to be real-time in the playback engine to respond to the ever-changing mix balance in the same way as with any real-time game audio engine.

I truly appreciate the aesthetic virtues of multi-channel delivery from a technical perspective and an immersive one when correctly set up. There are few ways to fully appreciate it as a consumer without buying a correctly set up listening room or VR system. Encoding to binaural for headphones has varied results whereas stereo and mono down mixing are clearly controllable. Maybe I am too old to see the future, but this future was invented decades before I was born and it still has not caught on with the general consumer in a purely listening context. Obviously as soon as this audio is linked to the visual, it becomes much bigger than its individual parts. In this regard every new format can become an opportunity for the engineer/entrepreneur who is quick to adapt to change and learn from the most recent development in the field.

For the consumer to effectively listen to an album that is mixed for Dolby Atmos, they will require a dedicated playback system to fully enjoy it, such as a 7.1.2 as a minimum in their home, which you would imagine should not be a high background noise environment. Equally, headphones using spatial audio would not be subject to high background noise if the user was trying to have an immersive experience. This means the concerns with dynamic range and translation in mastering should not apply. This potential useable wide dynamic range should be used to the full in the mixing process. The EQ balance should also be easy to manage because the Dolby mix room should sound excellent and be well controlled. All this means there is no need 'to master' the outcome in the same way with all commercial film outcomes mixed in post-production. You still require an engineer with excellent critical listening and sound balance skills, but not a mastering engineer directly. But this could be one and the same. Apple are keen to develop a multi-channel consumer base in this area though the Atmos mix will still be supplied with a stereo mix version as part of the delivery. This mix could be mastered for normal consumer use on Apple Music, but the spatial audio playback on this platform in binaural should be unique to the Atmos mix. Otherwise, the spatial audio encoder is just up-mixing stereo audio to create the impression of an immersive experience. Personally, I have not heard any music that ever sounded better in an up-mix than the stereo version using any multi-channel tool for this purpose.

Sequencing

When the mastering of a set of songs has been finished, the next stage is concerned with how they will be consumed together as a whole. The transitions between each song need to be appropriately edited to achieve the best musical transition. This means never applying fades or edits until all the audio processing is complete. This way all edits are committed at the highest resolution. Remember, a fade out is reducing in amplitude, and the bigger the resolution the better quality the fade will be. The larger the bit depth the better, making 32 bit renders of mastered audio more effective than 24 bit. I would always suggest working on the final HD render of the transfer path for each song and not trying to do this editing as a part of the transfer path outcome.

The transition between songs encompasses the whole of the audio from the start of the fade out (tail), through the silence after the fade end (gap or pause time), into the fade in (top), at the start of the following track until it reaches original file amplitude. Often the top is just a few samples or frames to avoid any potential click at the start of the audio. Having a straight cut can leave you open to this happening. There are no rules as to how long a gap should be, so our default response should be to listen rather than using some arbitrary unit like two seconds which does get mentioned as a standard (it is not). The correct transition time is the one that sounds correct for the music, whether it is crossfaded, or even a fade into a segue. This is a short musical piece or background noise to link the transition to the following music or outro. Sometimes people may also refer to this as a skit. Either way, continuous audio through the transition means you will need to decide on where the effective start of the next track is. At this point make sure the start of the track in the transient will be on a zero-axis cross point where the audio on both sides of the waveform are at zero. If this is not possible, you may need to address with a short fade in/out to avoid the click. I am personally not a fan of this and would rather work to move the marker away from the ideal start point to find an appropriate zero crossing or HPF the area to achieve less energy in the lows and minimise any potential click.

Fade types

There are three main fade types: linear (Lin), logarithmic (Log), exponential (Exp). If you combine a Log and Exp, one following the other, it will create a cosine (Cos) and the opposite way a sine (Sin).

Some DAWs have a more advanced fade curve editor (Pyramix/Sequoia) where a handle of the fade parameters can be outside the usual space dictated by the audio file. It is very useful, though you can equally create these shapes by using multiple crossfades with object editing, rendering silence on a file to increase the fade potential length.

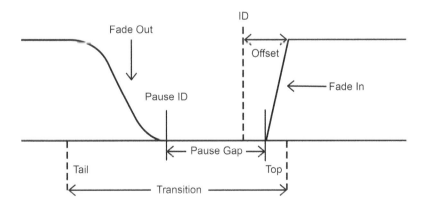

Figure 16.1 Sequencing transition parts names

227

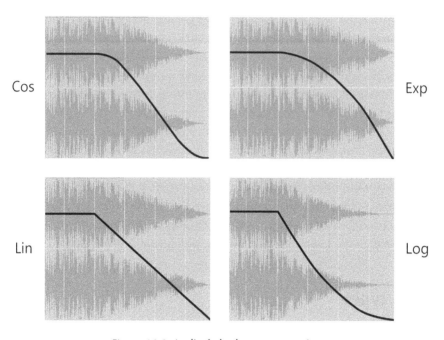

Figure 16.2 Audio fade shapes comparison

There are very few rules in audio, but there are some easily observed outcomes with fade curves. Firstly, our hearing is not linear. Making a linear fade is generally unhelpful, though it can go unnoticed when used in very short top and tail of a few frames, but this does not mean it is the most effective. Linear fades are great for equal gain crossfades, but otherwise it is not an option that will sound best by default. Our hearing is logarithmic so when continuous programme material is to be faded out, a Log (a classic musical fade) will sound best. Unless the fade is very long, over 10 seconds, a Cos sounds more effective as the initial drop takes longer and the Log aspect is in the second half of the fade. If the fade is over 20 seconds, it is likely any Log will sound laborious, as the main gain change happens quickly and the listener is left waiting an age for it to finally fade out. A Cos will address this more musically. If the audio already has a fade out, as that was how the audio was supplied, it is not ideal (see Chapter 4 'Looking to the mix'). Though if faced with this, when reducing the dynamic range overall and in-turn increasing perceived loudness, often the audio ends up with a small step drop off at the end of the fade, making it sound unnatural. In this case, when re-fading a fade, an Exp curve will smooth the outcome without removing too much of the original musical fade. As always, the correct response is to use our ears to make critical judgement based on what we hear and not what it looks like. When it sounds good with an effective transition for the music, that is the correct fade and curve.

There is always the possibility that even though the fade is correct in transition, noise in the top or tail becomes obtrusive in the outcome. Cedar Audio Auto Dehiss is the best tool for static broadband noise removal pre the fade. There are many noise reduction systems but Cedar Audio is streets ahead in my opinion, dynamic noise reduction is another option. Waves X-Noise is an effective frequency dependent downwards expander if not used aggressively. I would avoid using any noise print and stick to a broadband response in the fade transition to avoid too much change in the tone of the audio in the transition. You could make your own by applying an LPF to a copy of the fade object and crossfade between the original fade and the fade with the LPF making the LPF become more active as the fade progresses just ending in the object with the LPF. Also, you could automate the LPF, though I am never a fan of any automation in mastering, and always prefer to use audio and objects to achieve a definitive response from the DAW. Remember, automation is control data and not audio.

One important aspect to note is how to observe the transition by making sure you listen to enough of the song before the tail starts, this makes sure you will hear the flow correctly. When listening to the end of the fade out, in reality you are just observing the gap, listening to silence. The tendency is to reduce the silence until the transition is tight. When observing the context of transition and playing back the whole album in sequence, you will feel like all the tracks are too tight. Gaps are good, it gives the listener time to cleanse their audio palette before the next audio feast. Correct transitions can only be observed with enough of a context of the music around them.

Sequencing is a skill to help in your development. Analyse and assess albums you always enjoy listening to as a whole; there will be a reason apart from the songs you feel it musically works for you.

Audio offsets and framerate

Once the correct sequence and transitions have been created, the audio needs to be marked up for each track identification (ID). With CD Audio the track ID is where each track will start or musical movement can be auditioned from. In streaming or a download perspective, these split points are the playback start point when selecting a track/file. If the audio is hard edited to the start of the music (no offset), there is clearly the possibility for the audio playback to not activate absolutely on the start because the playback is not in sync with the stream of the audio. To avoid this potential click being evident, many systems ramp up the audio level on initial playback. This principle was first utilised with CD Audio and called the pause unmute time. This ramped up fade happened on the CD player on playback whenever the audio was skipped to another track ID to avoid clicks. This ramp is 12 frames in length with a frame rate of 75 compact disc minutes seconds frames (CDMSF). Most mastering engineers apply an offset to the ID in CD Audio to allow for this pause unmute somewhere between 12 and 25 frames, 160 to 333 ms. When download stores first started, they were using CD Audio as a source for ingestion, then offsets were part of the audio, which was helpful if the user downloaded to burn to a personal CD-r as the audio offset was embedded in the audio. The same premise still exists today. Different streaming platforms have the ramp ups on playback for the same rationale to avoid those clicks. If you want to stop the start of your master being chopped off by a pause unmute, you need to apply an offset.

Parts production

This is the action of making multiple versions of the master for delivery. Assuming the highest resolutions are being used our first output would be an HD master, 96kHz_24bit or 192kHz_24bit. In this final outcome there is no rationale for a 32-bit master because the audio is no longer going to be processed or increased in amplitude within a digital construct to take advantage of 32-bit resolution. Also, nearly all converters on playback will be a maximum of 24-bit. Following the HD masters – exports should be made for any lesser resolution formats required. This could be for video at 48kHz_24bit. Or CD Audio at 44.1kHz_16bit. The latter would be used to create a Disc Description Protocol (DDP) fileset for CD manufacture. But maybe the client requires an AB vinyl split with different formatting of the audio or same source master. This should be run at the HD resolution. Or for promo MP3s at 320kbps and so on, we refer to all these aspects as 'parts production'.

As a rule of thumb, start from the highest resolution and scale each out-come from there. For example, if our final master was 96khz_32bit, a 24 bit master would scale from that source as much as a requested 256kbps AAC delivery would be. This maintains the best outcome in translation of scaling the audio. It is important to be observant of TP changes because of sample rate conversion or lossy codec outcomes. When reducing bit depth in lossless formats, remember to apply appropriate dither. All this is a matter of testing to comprehend your own digital signal path and associated out-comes. Never assume anything and test, test, test. You will then know with confidence what will happen to your audio.

DDP was developed by Doug Carson & Associates Inc. as a carrier sup-ply format for CD manufacturing. Mastering engineers have been using this protocol ever since digital tape systems could be used, then optical and now data transfer is a definitive way to transfer and verify CD Audio formatting and most importantly the audio itself. This is just 44.1kHz_16bit audio in a different file carrier, it is still LPCM, like a wav file. Sonoris DDP creator is an excellent standalone tool for DDP creation, as are many mastering-oriented DAW such as Pyramix or Sequoia which can create these file sets with ease. There are some fundamental advantages to DDP, one is peo-ple cannot change it and it cannot be easily used unless you comprehend DDP. The files have a checksum, the same as a barcode does, making for a straightforward method of verifying that everything still is as it was, easily identifying where errors especially in transfer have been made.

Managing file delivery is constantly evolving, in the last decades the industry has finally moved from physical master delivery to purely via inter-net file transfer. DDP protocol has helped with its inherent ability to verify the data. Equally important have been zip formats, not to compress, but to interleave all the files and folders as one. A single master file, which if corrupt in transfer or copying simply will not open, is another layer of verification. You might hear some discussion around zipping and reduction in quality. Simple sum/difference testing can verify if a digital file is one for one. If people ignore this and insist there is a difference in playback, I would surmise they are hearing jitter in their playback monitoring system and that is the source of their error, not the audio itself. Investigating zip file type compression, you would find it is possible to not apply any file size change, but simply let it act as a wrapper for the files. When delivering any file set for ingestion, I would always zip with associate track sheets or table of contents.

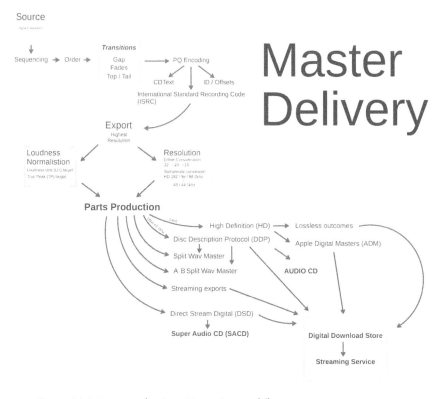

Figure 16.3 Parts production: Mastering workflow to consumer outcomes

Recall

To achieve reprints of any master for whatever reason, there is a need to have a methodology for recall. Even when working purely in the box, there is a need to be considerate about archiving and saving each individual master outcome in a way that it can clearly be recalled. When coming back, it is important to understand what was done and why it happened and how that was used in the production of the final master. Personally, I track all jobs with a unique job number, meaning all aspects are associated with this to locate and recall any hardware or software that may be saved on a separate system to the DAWs in use. All this can be easily found and archived together for later recall. When file naming, any notable action I take on a given master is added to the file name and version number used for each of these to create a hierarchical tree of development, much in the same way mixer automation systems used to be used. These are clear roots and branches to the final out-come, meaning it can trace and change and update. Most importantly you can be confident the recall being used is associated with the correct masters. I cannot overstate how important this is if you ever intend to work commercially in the audio industry. It is not a choice, but a requirement.

Variants

The following are all possible extra prints that could be required for a single or a whole album or as HD, CD and/or lossy qualities.

- Cleans (expletives removed).
- Instrumentals (no vocals).
- Radio edits (shorter versions for radio and TV).
- Sync/PA (no main vocal but all backing vocals, sometimes called a 'sing back').
- Beds (broken down versions of the song, only drums and bass together and so on).

These prints should be made once the main masters have been signed off. Accurate recall is an absolute necessity to achieve this, as is the accuracy of the original mix prints. I would recommend testing variant mixes in sum/difference to check validity before running any print. I would equally test the prints against the main master as a final check once complete. These are run in real time verifying the validity of the printed audio but it still needs to be checked again after editing up and rendering. All this takes time. Variants are not something to offer up as a free addition included in mastering. Parts production takes up a considerable amount of time in any commercial album mastering or single production. Make sure you charge for your time, listening skill and knowledge.

Verifying audio masters

The last and most important part of a mastering engineer's job is listening to the final master for production. If failing to observe the final outcome for errors or anomalies, at some point errors will be missed. Having listened to thousands of final outcomes, I found that aspects missed come to light more often than you would initially think considering the amount of critical listening that has taken place before this final point. Sometimes these aspects are noticed as the context is now finally listening to the album as a sequenced musical unit, but there can be more serious glitches or errors in the audio that have happened in the final stages of rendering or have just been missed. Even though I have been working in real time throughout the whole process, and listened several times to individual songs in their entirety, the context of this final listen gives a different perspective. Make one last final check to make sure everything is 110% perfect. Your final outcome is the version that will be put into production. There is no going back, and you have to ultimately shoulder that responsibility.

17   What's next?

The right way, the wrong way

There are three aspects that in my opinion make it difficult to achieve a positive outcome when mastering. The primary factor is equal loudness. Lack of adherence will mean incorrect or poor decisions will be made some, if not the majority, of the time. This means consistency in delivery cannot be achieved because decisions are flawed; the listening conducted is bias by loudness. The second is listening level, if the ideal of around 83dB is not being observed, or an inconsistent listening level is used, inherently incoherent decisions will be made some of the time. This lack of adherence to basic good engineering practice is just making life needlessly difficult. Though equally, it takes effort and self-regulation to stay on top of these outcomes to maintain a constant listening level. A clear focus on the correct techniques to achieve great outcomes with yours and others' art must be constantly maintained. The third aspect is not listening to the product output printed. This has become possible as processing has become faster offline. The number of errors I have observed in mixes supplied by engineers because they have not listened to their final output render has grown substantially over the last couple of decades. Trusting offline processes will lead to errors, it is inevitable. Errors happen in real-time rendering/printing, but the difference is the engineer is listening to the recording file being printed. As any issue can be heard, it can be correctly re-printed. If rendered offline, the next action must be to listen in full to the output file before sharing it with the client. This is a simple rule, never to be broken.

DOI: 10.4324/9781003329251-18

Managing expectations

Mastering is an art; it takes considerable time to develop confidence in your own approach. Much of this is developed through evaluation and self-reflection. You are not going to achieve your optimum potential overnight, but you can make the best outcome possible at this current juncture based on effective appraisal of the products you create. It means you will have delivered the best you can. This is a good outcome. In the future you will be able to make more effective decisions, but this does not mean what you create today is flawed. Confidence in your own output is critical. This comes from the knowledge that you have genuinely improved the outcome of the project because it has been evaluated effectively. This goes the other way around in dealings with the client. Mastering engineers are not miracle makers; the song, recording and mix is what it is. This does not mean it can be turned into something it is not. Working with the mix engineer to improve outcome will always be helpful, but also do not be afraid to turn away mixes if you know it is not going to achieve a useful outcome. Sometimes, it just needs more work, even if the production team thinks it is ready. The more you work with repeat clients, the more they will value this honesty in appraisal. I have never come across an artist, client, label or production team that did not want to get the best out of the project as a whole. In working with someone's art, time, knowledge and skills are being invested to achieve the most effective outcome possible. Do not be afraid to have conversations about outcomes across the production process. If approached with humility and not criticism, everyone benefits in the final outcome. After all, that is what everyone involved in the project is judged upon.

Adding value, being valued

Working in the music industry is challenging. The main advantage you have is skill, but there are plenty of people offering something for nothing in the world. If you are at a stage where you are investing time expanding your knowledge of your art, then inherently you have already achieved a positive level of skill. You should charge for it, you have value. Clients who want something for nothing do not value you or the music they are involved with. These are the types of potential clients you should avoid. Those who are willing to pay, even if this is modest because of their budgets, to them, this is a lot and thus has added value. Clients who value you are also likely to value the art they are involved with. These are the clients to focus time and energy on rather than those that can pay, but choose not to. No matter how influential they may appear to be, in the long term it will not be worth the effort on your part. Networks are built from like-minded people being creative. Make your positives be positive. Nobody has ever made a sustained and successful career based on working for free.

Belief in your critical listening

If you are going to achieve excellent outcomes, you need to trust your ears. If you know something does not sound right, that is because it is not correct. You may not be able to find what is awry initially, but if you hear it, it must be there. Persistence and testing are required until you work out the issue. The other aspect here is honesty in analysing the final master. Absolute adherence to equal loudness and relative loudness normalisation is critical. Comparison against the source mix will allow you to make a realistic appraisal of the translation of the mix to the master you have created. You must clearly observe that the master has improved in all aspects during evaluation or you will have to return to the transfer path and reassess. If you do not do this, you are making assumptions and not achieving the best outcome for the audio or your client. Improvements are only realised with critical reflection of our own practice. This fundamentally requires honest self-reflection.

Test, test, test

When the intent is to be truly critical in our application of tools, there is no rationale not to thoroughly test our processing, and the methodology of the process. If rigour is applied in this, our application of method and engagement with a given processor will inherently be more effective. These small gains in focus of tool use all add up to much larger gains in the bigger picture of your mastering output. These aesthetic changes are the difference between a master sounding smooth, edgy or impactful on multiple playback systems. This is the timbre of a master.

Nobody knows the answer but you

To me, one of the most interesting things about engagement in any art form is the lack of constriction. Who is to say where the boundary is, or how it should be set, though the consumer is the one that in the end judges the success of how our output is consumed or adopted. As the mastering engineer, you are only one part of the production chain. In the end of it all, the song and performance are clearly the most important and powerful input. But you must be the judge in relation to the quality of your output. You are your own moderator, and will be judged by others on that output and not what gear you used or techniques engaged with, the consumer appraisal is the most telling of all. They just like or dislike it, you cannot get a more honest opinion than that.

A note on artificial intelligence in mastering

In considering the methodologies discussed throughout this book, the observations and evaluations of the approach to mastering are quite logical. A list of aspects viewed in analysis of mixes regularly could be linked to a process and method of pathing. A list of do's and don'ts around how differing tools interact could be written. Even formulation of our analysis into a given path would make most things at input sound better at the output. Better, but that does not mean correct or aesthetically interesting in any way. Controlling the levels of the input by standardisation into the signal path would improve the outcome helping to manage thresholds on dynamics and Fast Fourier Transformation (FFT) can manage 'ideal' EQ curves. If the genre was known, the chain could be tweaked to generalise the main tropes in that music arena. In short, make an automatic mastering path with a user selectable genre input, taste controls and level of output definition.

This is not intelligence as such, but it is using data sets and medians to achieve average responses to an infinite number of potential inputs. The term artificial intelligence (AI) has been promoted in the mastering arena in recent years. Systems seem to use logical algorithmic consideration based on semantic measurements of audio and average responses from collected data and most importantly user input to tune the outcome. You could say machine learning is another way to describe parts of this. Either way, it is still reliant on the user's initial input and response to output to achieve something that is acceptable to the client. It is certainly not a programme thinking about how the music sounds. It might be measuring it, by using semantic audio methods. Though currently there is not even a level meter that can measure loudness absolutely accurately from one piece of music to the next in our perception of musical equal loudness, because it is measuring the signal and frequency and not listening to the music.

All of our logical steps above will not work effectively if listening is not the focus in the first place. Even in the simple act of achieving equal loudness, to make true comparison between a mix and the reference is very complex. But to a trained engineer, it is this easy and intuitive human action of listening that is the most critical in all of the mastering process. Put simply, if we cannot hear it, we cannot change it! This goes for any automatic system that may be devised, it is the listening that needs to be the key. When automatic audio systems can listen and interpret musical context, at that point, something that could be described as intelligent automatic mastering could possibly exist.

I do not foresee this happening anytime soon, but that does not mean these current systems do not have their place in the consumer market place. If something can improve an audio outcome, surely that is a good thing. Maybe the way it is marketed makes it harder to interpret and take seriously in a professional context. If you are not an expert in a field and are using AI to deliver an outcome, what you receive from the AI system may seem to be professional to you. But to an expert, it has clear flaws and does not translate or scan correctly. It is all about perspective, and from this point of view it is clear to see how some music makers find AI mastering a great tool to improve their outcomes.

When musicians progress in developing their art, a point may come where they start to hear the flaws and will seek out help, that is where professionals step in. This progression is nothing new in music creation. Technology and its developments should be embraced and supported to evolve. After all, where would music be now if everyone decided to join the Luddites and destroyed the first compressors! It was somebody's job to control the level of the broadcast manually before its invention. Technology took their job, or freed them from what became a mundane audio task; it all depends on your point of view. I personally have embraced technology over the years and will continue to as it develops to help make the music of the future sound better.

In this context, I foresee plugins with semantic audio responses becoming more widely used professionally to enhance the response of an excellent sounding plugin. The potential in this area is vast, but the key to their success is in the interfacing with the user. If a professional engineer can use it to experience a more effective outcome with less contact time with the audio, I am sure it would become a tool of choice. Put simply, I think the future is still with the artist and engineer doing the listening. The tools used now and as in the past are just facilitating our artistic responses and not creating it.

Perspectives with multi-channel audio

Music does not need to be more than mono to be enjoyed. It does not need more than one speaker, and that speaker does not even need to be good quality. As the previous consumer formats have taught us, multi-channel systems are not needed by the average consumer to get full enjoyment out of their music, but a small minority might buy into the requirements. The majority just want to listen to the music. The rationale is different for cinematic sound or multi-channel sound design, especially in gaming/VR where the sonics are part of the visual landscape. Overall, the consumer dictates the playback outcomes, not the technology or our wants as audio engineers or artists. Focus needs to be on making our output the best quality possible for the simplest format translation from mono and stereo, and it will translate on all current and future systems. We can deal with other multi-channel formats in whatever configuration is required, as and when, but mono is the queen and king.

That said, I do not mean to sound negative toward this exciting area of creativity. I have an open heart towards multi-channel audio, it is a wonderful experience to work with sonically. Professionally the opportunities are limited in a purely audio arena, but if you find your niche, especially with a genre, there is a world of opportunities to embrace. For further reading in this area, Darcy Proper and Thor Legvold submitted an informative chapter for Mastering In Music [11] published by Routledge.

In fully understanding the translation/downmix from stereo to mono, all other multi-channel systems can be understood. LR is MS, this is one dimension, add another MS front/back, another up/down and the outcome is basically B-format, WXYZ. Ambisonics or true phase coherent three-dimensional audio sound capture and playback are used in VR/gaming audio engines. Everything in audio eventually comes back around to amplitude and phase. Understand that fully and you can interact with any audio outcome proficiently and professionally applying knowledge and skill to the challenges presented.

What is the future?

For me as an engineer, the tools I use have changed constantly over time. As a theme, the equipment I used a decade ago I no longer use, other than a few pieces because the majority of audio equipment has been sonically surpassed or user interface has been improved. The march of technology is relentless and on an exponential development curve, meaning aspects used today will not be the same as those used in the near future. But the physics of sound does not change, and dynamic principles have not wavered since the invention of the first amplitude control. If you are developing skills, focus on the 'why', and the principles of use and do not just engage in repetitive interaction with a specific plugin or hardware unit. Learning to be proficient with only one DAW is unhelpful, learning how to interact with a DAW and how it functions and why it works that way it does means you can use any DAW or analogue mixer or hybrid device. The skills learned need to be transferable. As technology develops an engineer can move seamlessly with these changes. picking and choosing tools from the past and present that work best for their audio needs of their current paradigm. In the same way that signal path was once purely analogue and is now mainly digital, the realm may have changed, but all the principles in the use of those signal paths have not.

Your future development in audio will be led by your observation of good practice, much of which is explained in this book. By embracing critical thinking and evaluation, you will become a more effective audio engineer in whatever aspect of the industry you engage in. Learning gained in any area of audio is tangible in use elsewhere. After all, signal path is just signal path wherever the audio is.

References

[1] Fletcher, H. and Munson, W.A. (1933). Loudness, Its Definition, Measurement and Calculation. *The Journal of the Acoustical Society of America*, 5(2), pp. 82–108.

[2] EBU (2000). Recommendation R68–2000: *Alignment level in digital audio production equipment and in digital audio recorders*, EBU, Geneva.

[3] "SMPTE Recommended Practice - Reference Levels for Analog and Digital Audio Systems," in RP 155:2014, pp. 1–7, 29 Dec. 2014, doi: 10.5594/SMPTE. RP155.2014.

[4] Katz, B. (2000). Integrated Approach to Metering, Monitoring, and Leveling Practices, Part 1: Two-Channel Metering, 48(9), pp. 800–809.

[5] "SMPTE Standard - Motion-Pictures — Dubbing Theaters, Review Rooms and Indoor Theaters — B-Chain Electroacoustic Response," in ST 202:2010, pp. 1–9, 20 Oct. 2010, doi: 10.5594/SMPTE.ST202.2010.

[6] Nyquist, H. (1924). Certain Factors Affecting Telegraph Speed1. *Bell System Technical Journal*, 3(2), pp. 324–346.

[7] Watkinson, J. (1989, Sep.). "The AES/EBU Digital Audio Interface," AES Conference: UK 3rd Conference: AES/EBU Interface.

[8] AES67–2013 (2013, Sep. 11). "AES standard for audio applications of networks - High-performance streaming audio-over-IP interoperability," Audio Engineering Society.

[9] ITU-R Rec. ITU-R BS.1770–4 (2015). "Algorithms to measure audio programme loudness and truepeak audio level," International Telecommunications Union.

[10] EBU. (2014). "EBU Technical Recommendation R128 - Loudness Normalisation and Permitted Maximum Level of Audio Signals," European Broadcasting Union.

[11] Braddock, J.P., Hepworth-Sawyer, R., Hodgson, J., Shelvock, M. and Toulson, R. (2021). *Mastering in music*. London: Routledge.

Index